MW01673590

Why Knot?

A Personal Quest

Richard Seltzer

BookLocker
Trenton, Georgia

Print ISBN: 978-1-958878-47-7
Ebook ISBN: 979-8-88531-405-3

Published by BookLocker.com, Inc., Trenton, Georgia.

Printed on acid-free paper.

BookLocker.com, Inc.
2023

First Edition

Library of Congress Cataloguing in Publication Data
Seltzer, Richard
Why Knot? A Personal Quest by Richard Seltzer
Library of Congress Control Number: 2022922373

Dedication

For Nancy, for pushing me to say what I mean simply and clearly,

Acknowledgments

To my children (Bob, Heather, Mike, and Tim) for teaching me so much from the time they were very young.

To Rex Sexton and his widow Rochelle for decades of support and inspiration.

To Claude Thau and David Lupher for many contentious discussions that helped me fine-tune and clarify my thoughts.

To my readers at Medium.com for their feedback and encouragement. To Gabi Coatsworth for her writers' meetups that made me aware of today's publishing options.

And to all of the folks at Booklocker who worked on this book, in particular Angela Hoy the owner and Todd Engel the creative cover designer.

Foreword

I don't think *outside* the box. The very notion of "the box" is an illusion that limits the range of topics, squashes curiosity and creativity, and precludes innovative solutions. Please join me on my journey of exploration.

Dream November 26, 2022

During a press interview, I was asked about my education and IQ. How could I write about such a wide range of subjects?

I replied that that was irrelevant. I dared to do what I did because of the confidence my parents gave me as a child. They taught me to question assumptions and challenge authority, to respect common sense and to find my own answers in my own ways.

Interpretation

I don't think they did that deliberately. I was their only child. To them, from birth, I was a miracle. So, of course, I challenged everything they believed in. I was not an easy child or an easy student.

When I was 12, I tried to write my own version of the Trojan War story, in rhyming couplets. When I was 13, I brought to Sunday school class a list of my beliefs, the core of my personal religion. When I was 15, I started keeping a list of every book I read and added to that what I remembered reading starting at age 12. I've kept adding to that list ever since. It now includes over 3700 books.

My beliefs have changed radically over the years, and what I wrote when I was 12 and 13 looks silly today. But my self-confident style in dealing with the unknown and the unknowable is much the same.

Epigraphs

"...we suspect our instruments. We have learned that we do not see directly, but mediately, and that we have no means of correcting these colored and distorting lenses which we are, or of computing the amount of their errors. Perhaps these subject lenses have a creative power..."

— from "Experience" in Emerson's *Essays*

"The soul is a prism
That casts rainbows
From heaven"

— from *Skeleton Key* by Rex Sexton

Table of Contents

Part 1:
Big Questions

Toward What Goal?

I need to know who I am, why I am, and how my life might matter. The answers offered by religion feel insufficient. Scientific knowledge has advanced so far that it is beyond the understanding of laymen. I would like to participate in the endeavor of scientific discovery and contribute. But the advancement of science will not end in my lifetime and will probably never end. And I need answers that make sense here and now.

I do not adhere to any organized religion or established set of beliefs. But I am not an agnostic. Rather, I am a *seeker*.

Life has meaning. Perhaps that meaning lies in interconnectedness and relationships, to be discovered by intuition rather than reason, in flashes of insight.

There is Everything We Can Do

One of the major discoveries of the twentieth century was recognition of the limits of human knowledge. That principle, seemingly an admission of weakness, can be a source of strength.

The experiments we perform to understand the world change it, hence limiting to our knowledge, leaving us with an underlying uncertainty. But by that same principle, all of us, regardless of our strength or intelligence or social status, change the world by what we do. If and when people coordinate what they do and act in harmony with one another, they can make major changes.

That realization is important as we face critical issues such as pandemics and global warming. Individually, we are helpless in the face of complex and pressing dangers; we can do nothing. But working together we have enormous power.

What's the *It*?

To the question of life and death, my friend says, "It's over when it's over." She means there's nothing after death. There is no soul. There is no God.

I wonder — what's the *it*?

Descartes' *I think therefore I am* presumes there is an *I* and that we know what *I* is.

Similarly, *It's over* presumes there's an *it* and that we know what *it* is.

I don't expect answers. I read and write in hopes of asking ever more intriguing questions.

Getting Personal

In my reading, I sometimes come across passages that feel right, not as discoveries, but rather as clear and cogent expressions of what I believed before, stimulating me to see my own thoughts from different angles.

For instance, consider a passage from Boethius, who wrote in the sixth century. In prison, awaiting execution at the random whim of King Theodoric of Italy, as he tried to make sense of life, he concluded that infinity, eternity, and chance reduce everything to insignificance.

I am undaunted by such considerations. I hope to arrive at personal answers that help me to deal with day-to-day reality. I strive to find truth and meaning in the context of my everyday life. I'm on a personal quest to try to understand what matters to me as an individual, living here and now.

Science?

Matter and energy are neither created nor destroyed.

But life is both created and destroyed.

Is life neither matter nor energy?

Then, what is it?

Where There's a Will

We equate consciousness with rational thought, and we can correlate thought with brain activity. When there's no brain activity, we presume there is no thought, and we say that person is *brain dead.*

But we often act without thinking. We also can make a conscious decision to do one thing, but do something else, even the opposite, surprising ourselves. In other words, the will, though associated with thought and a subject of thought, is separate from it.

Is the brain necessarily the seat of the will?

Language associates the will with emotion and intuition, and suggests that the will is centered somewhere other than in the brain, for instance, in the heart or gut. Language also associates the will with such concepts as "soul," "self," "spirit," and "life force."

Does the will necessarily cease when thought ceases? Might someone who is declared brain dead still have the will to live?

In language as well as in religion and myth, the soul or spirit is separate from the body and persists even when the body dies. So why presume that soul/self/spirit/will has the same physical location in the body as thought, and that they all end at the same time?

Listening to Life with a Tin Ear

I used to envy people born with perfect pitch. Unlike me, they could appreciate music to its fullest. I couldn't tell if a piano was out of tune, nor distinguish great from mediocre performances. But now I've reached an age when instead of regretting my limitations, I relish them.

Perfect pitch is a curse, a tin ear a blessing. To someone with perfect pitch, anything less than a perfect performance is painful to hear. He or she can appreciate subtleties beyond my ken, but probably can't enjoy what passes for music for the rest of us.

I can appreciate a flawed performance on a piano that's out of tune. I can enjoy sing-alongs and amateur singing and karaoke and informal musical events. I can delight in whistling while I walk. My opportunities for musical pleasure are far greater because of my tin ears.

Similarly, I can savor ideas that aren't fully developed. I can enjoy a story, a book, a movie that is good but not great. I have everyday, non-professional expectations.

I'd rather explore many subjects than devote myself to one narrow field and never arrive at complete knowledge of that field. The world is far too complex to understand in detail.

I know I can never get definitive answers to the big questions, nor can anyone. Instead, I want to arrive at personal answers that make sense on the scale of where and when I live, despite my awareness of the vastness of infinity and eternity. I need lenses that help me look at the world with a perspective of immediacy, from the context of daily life.

Let's enjoy what we can know. Let's enjoy life as best we can, glorying in the imperfection of our tin ears.

Observations on the Meaning of Meaning from Practicing the Piano

As I was practicing the piano this morning (as a talentless amateur), my eyes focused on the sheet music, only occasionally glimpsing my fingers on the keys. For pieces that I'm familiar with, my fingers seem to have a memory of their own. They remember not the keys but the intervals and the sequences, while my eyes focus on the visual patterns of the notation.

At the beginning of a piece, I look at the sheet music to determine the key (defined by sharps and flats) and at my hands to properly position them for the opening notes. As I play, I compare the sounds I hear with the sounds I expect to hear. And I'm also alert to instances of dissonance, which, independent of my memory of the tune, signal that I've struck a wrong key.

If I'm playing a song, I'm also aware of the words that go with the notes. If I were to recite them without playing the piano, even if I did not sing the words, I'd be conscious of the tune.

This observation reminds me that language can be associated with sound in complex, multilayered ways. I am aware not just of the semantic meaning of the words, but also of the tone, the rhythm, and the patterns of speech that define a local accent or dialect. I can guess the speaker's place of origin and state of mind. And, simply from the sound, I can detect the identity and sincerity of the speaker—whether the words mean what they seem to say or maybe the opposite.

Handwriting, too, conveys more than the semantic meaning of the letters and words. The style can be so unique as to reliably indicate who wrote it. And sometimes, when you know the writer well, you can tell his or her state of mind at the time of writing.

In touch typing, I don't think of the individual letters and don't worry about the spelling. Rather, I think and type complete words and phrases, without looking. My fingers, placed in the right position, remember what they need to do, apart from conscious thought. Only when I proofread do I discover the spelling mistakes my fingers made, often substituting one word for another, rather than jumbling letters mechanically. I expect that the hands and fingers of accomplished pianists remember large chunks of music rather than individual notes.

My consciousness seems to act like the conductor of an orchestra, setting the tempo and helping coordinate multiple streams of activity. But my consciousness has less control than a conductor. Many bodily functions, like breathing, operate semi-autonomously; and others, like the beating of the heart, function on their own.

Among all those layers of thought, where am *I*? Who am *I*?

My focus shifts depending on what I am doing and how much attention it requires. Some functions know what to do and how to do it from habit and training. Some, like walking and talking and chewing gum, can go on at the same time without needing my attention. Some skills, like riding a bike, that I may not have used for decades, I can do again without effort or thought. Speaking a foreign language and playing the piano require refresher work, but the initial training makes relearning much easier than learning from scratch.

In other words, we think in layers, and what we call *consciousness* is just one of those many layers.

Highly Unlikely

The chains of events that shaped my life were highly unlikely — one coincidence happening after another. If the chain had not unfolded just the way it did, everything would have turned out differently.

If you ever fell in love, think about the events leading up to that moment. After the fact, the events feel inevitable. It's difficult to imagine how your life could have gone if they had not occurred when and how they did. All the pieces fell into place miraculously.

The more you know about an event, the more unique it seems. To you, the events of your own life seem like the result of an extraordinary series of coincidences. But an outside observer, with far less information, would view those same events separately instead of as a sequence. Separately, they would appear ordinary and expected and not coincidental.

Every time you toss a coin, the probability of heads is 50%, regardless of the results of previous tosses. But a long chain of events, such as heads, heads, tails, heads, tails, tails, tails, defies analysis. Only when you isolate a variable and simplify the context with a generalized perspective do the laws of probability apply.

According to Bernoulli's Law, one of the basic principles of probability, it's possible to accurately predict the average outcome of many similar events, but it's impossible to predict, with certainty, the outcome of any single event. In other words, the more you know about a specific event and the chain of circumstances that led to it, the more unique that event appears. Hence, every moment of your life is miraculous.

The Abraham Effect: The Future of Humanity Depends on You

Just as the entire Jewish people are descended from Abraham, the people who inhabit Earth a thousand years from now may all be descended from you.

You have two parents, four grandparents, eight great-grandparents. The number of your ancestors doubles with every generation. Counting backwards 1000 years, about 36 generations ago, you had about 69 billion ancestors. That's two to the power of thirty-six. But at that time, there were only about 50 million people alive in Europe. Everyone of European descent alive today is a cousin of everyone else, and all of us have the genes of a few people who were alive then.

Fast forward a thousand years, taking into account that people are far more mobile today than they were a thousand years ago. In the year 3000, if the human race survives that long, every human being alive on Earth and on other planets will be a descendant of a few people who are alive today. If you are a parent, there's a chance that everyone alive a thousand years from now will have genes that passed through you.

That is an awesome responsibility. Be careful. Be proud. The future of the human race depends on you.

Our Time

You perceive time very differently than today's computers record it. The amount of information your brain stores in an interval of physical time affects the subjective time that you perceive. Under stress, your brain processes and stores far more data, far more quickly than normal. There are limits to what can be stored in short-term memory. In moments of crisis that limit is broken, and short-term spills over to long-term, and the mass of data that is perceived gets imprinted in long-term memory and takes up far more memory capacity than is normal.

In popular wisdom, when you are near death, your entire life flashes before your eyes. In the moments before death, that flash might happen repeatedly, as time subjectively expands, in a variant of Zeno's Paradox. Just as Achilles never catches up with the turtle, you, subjectively, never die. Death is the limit that you approach but never reach.

To anyone else, your timeline ends. But from your perspective, you keep approaching death, forever: you never die.

Time Dilation

In the summer of 2012, driving back to Boston from Cape Cod, I came close to death.

I was alone, driving a van packed tight with stuff we had brought to the Cape for a two-week vacation. My wife, Barb, was cleaning the cabin and would be following in our other car in about an hour.

Three miles from the Sagamore Bridge over the Cape Cod Canal, I realized that my brakes didn't work.

The traffic around me was travelling at 60 miles an hour. The distance between me and the car in front of me was a car length. The car behind me was also a car length away. There were cars to the right of me as far as the eye could see. To the left of me there was a metal barrier.

The car ahead of me slowed. I gently tapped my brakes. Nothing happened. I tapped again quickly. Nothing.

Now I was just a couple feet away from the car in front of me. I stomped down on the brake pedal, and the pedal went all the way to the floor with no resistance, no change in speed.

Fortunately, the car ahead accelerated.

We were going downhill. I was coasting.

Options rushed through my mind.

I tried to downshift, but the gears were locked.

I considered using the emergency parking brake. But if I stopped suddenly, the car behind me would slam into me and I'd end up in a pile-up.

I considered turning off the ignition. But the van I was driving had power steering. If the engine turned off, the power steering would shut off as well.

After what felt like an hour but probably was less than a minute, around a curve, the hill ended, and I found myself on a slight incline. Then, a grass median strip opened up to my left. I turned left onto the grass, and the car started slowing down. In what felt like another hour but was only a few seconds, the car came to a stop, a hundred yards from the bridge.

My heart was racing. I saw the van, the grass, the road, the traffic, the beautiful blue sky with a clarity I had never seen before. My mind was muddled, but I was feeling ever so high, so relieved. I was alive.

I had never felt so alive.

The Time Between Time

A series of still photos or drawings viewed in rapid succession looks like natural movement. The faster the sequence, the smoother and more natural the motion seems.

The human eye and human brain evolved with this ability to convert a sequence of still images into the perception of motion. What was the survival benefit of that?

Perhaps reality is discontinuous, like a series of still shots; and the ability to perceive it as continuous provides practical benefits, as does the ability to perceive matter as solid and continuous rather than as atoms and force fields, mostly empty space.

We might ask if reality consists of smooth, continuous changes or discrete changes. If discrete changes were small enough, we wouldn't perceive them. Our perception isn't fine-tuned enough to resolve this question, any more than it allows us to notice the breaks between frames of a movie.

We can make machines that perceive and record far more accurately than our all-too-human senses and brain. But, today, the machines that we rely on to extend our sensory processing and memory capabilities are all digital — based on discrete choices — yes or no, one or zero. In other words, they have a built-in bias. Hence the resolution of this question is beyond their ability as well.

Is Reality Discontinuous?

When Isaac Newton explained the laws of gravity, he presumed that those laws would apply not just for Earth or our solar system or the observable stars — but everywhere.

A hundred years ago, in *A Pluralistic Universe*, William James came to a different conclusion. He speculated that reality isn't necessarily neat or logical or predictable. Beauty and simplicity are not synonymous with truth. Rather, the world we live in is messy and full of surprises, and the farther from Earth, the more likely the surprises. He believed that scientists should continuously test their assumptions.

Recent books such as *The Elegant Universe* and *The Fabric of the Cosmos* both by Brian Greene, *Warped Passages* by Lisa Randall, and *Parallel Worlds* by Michio Kaku explain the many flavors of string theory, which is a possible successor to quantum theory, which was the successor to relativity, which was the successor to Newtonian physics. These books describe a variety of ways to explain the data that scientists have gathered — multiple universes, multiple dimensions, dark matter, dark energy, and negative gravity. They all presume that we can understand realms of being that are far beyond our normal experience.

But our ability to make sense of the world evolved in this world. Our senses and our reasoning power are adequate for everyday life. But I believe we are not equipped to understand what happens on scales smaller than an atom or larger than a galaxy, much less in multiple universes.

There is no reason to presume that the universe, viewed through our limited and flawed senses, is simple and logical enough for us to understand it. Rather, the universe may be complex and discontinuous. Natural laws that apply in our solar system and in our galaxy may not apply elsewhere or may not be stable; and if natural laws change, they may not change in predictable ways.

For instance, consider Hubble's Law. As Wikipedia explains, "...the redshift in light coming from distant galaxies is proportional to their distance." Our calculations of the distances from Earth to stars and galaxies depend on that principle, presuming that the same natural laws that apply here also apply thousands, millions, and even billions of light years away. That assumption has mind-boggling consequences. If there are, in fact, discontinuities in reality and variations in natural laws beyond our galaxy, then what scientists have concluded about the size and nature and past and future of this universe, much less other universes, is in serious doubt.

We should consider the possibility that reality is messy, and that complex answers may sometimes prove more useful and suggestive than simple ones. Maybe there are two or more realities unfolding in parallel. Blink and you switch to a different life.

The simplest answer may not be the only answer or the most accurate or the most interesting. And the universe is very interesting.

Going Beyond the Limitations of Science

Science progresses by testing educated guesses — hypotheses. But our hypotheses depend on expectations based on our previous knowledge and cultural bias.

We face the same limitations in everyday life. We filter what we see based on what we expect to see. We ignore anything seriously out of the range of our expectations. If we don't ask the right questions, we don't get the right answers. And as human beings, we have a limited range of hypotheses that we are likely to consider plausible. Intuition and thinking-outside-the-box can expand that range, but not by much.

Today, computer simulation is used in conjunction with physical experiments to generate hypotheses and then test them. But such simulation typically stays within the range of human expectations.

To overcome that limitation, we need programs which generate hypotheses that are implausible and would not otherwise be considered; programs that come up with complex and improbable ideas and ways to test them. Such hypotheses could lead to experiments that record and help interpret data that would otherwise be ignored.

In the Middle Ages, the rule of thumb known as *Occam's Razor* was important in setting the stage for scientific advancement. "One should not increase, beyond what is necessary, the number of entities required to explain anything." That rule made practical sense because humans have limited time and limited brain power — focus your research on the most likely explanations. In today's vernacular, that's expressed by the acronym *KISS* — "keep it simple stupid."

But today's computers can deal with far more variables than humans can. They can calculate much larger trees of causation and hence can identify multiple explanations for the same event, all valid from different perspectives, and perhaps each leading to different long-term consequences. It's time to move beyond Occam's Razor, to expand the range of our research to deal with the complex, the unlikely, the redundant, and even the totally outlandish, admitting the possibility that truth might be messy rather than systematic and beautiful.

An Ultimate Unit of Space and the Need for a New Calculus

When I was reading *The Fabric of the Cosmos* by Brian Greene, which explains superstring theory for the masses, I was also reading *Quicksilver* by Neal Stephenson, an historical novel with Sir Isaac Newton as a character. On p. 670 of *Quicksilver*, one of the characters challenges a basic concept of calculus. He asks, "What happens then if we continue subdividing?... Is it the same all the way down? Or is it the case that something happens eventually, that we reach a place where no further subdivision is possible, where fundamental properties of Creation are brought into play?" The character is contrasting Newton's notion of infinite subdivision, with other concepts of the world in which there is a natural limit to such subdivision.

There appears to be a contradiction between superstring theory, which postulates an ultimate unit of length, and the assumption of calculus that space is infinitely divisible.

I sent an email to Brian Greene, wondering if fundamental concepts and procedures of calculus need to be refined to take this ultimate unit of length into account.

He was kind enough to reply. "In fact," he wrote, "that is just what we are working on today. The notion that the usual procedures of calculus are only relevant on length scales larger than some lower limit — we are trying to piece together the new procedures that take over."

Does Dark Matter?

We're told that dark matter and dark energy account for more than 95% of all there is in the universe. Ordinary matter amounts to less than 5%. The exact numbers change with new scientific advances, but the overwhelming dominance of the dark over the ordinary remains constant.

You can't see dark matter. You can't feel it or smell it or interact with it in any way. In aggregate, dark matter and dark energy account for the gravitational force that is necessary for equations fundamental to our understanding of the physical world. Basically, dark matter and dark energy are a fudge factor. If we want to believe that we understand the physical world and that the physical laws which hold true in our solar system and our galaxy also hold true billions of light years away, and that we can look back 14 billion years and ahead billions of years and understand what was happening and what will happen, then we have to believe in dark matter and dark energy.

But concepts like spirit, soul, and self are matters of mystical speculation.

Defining Action

After World War II, authors like Sartre and Camus reacted against abstract philosophy that neglected the immediacy, emotion, and empathy of everyday life. They blamed abstract, dehumanized thinking for the horrors of the Third Reich. Instead, they harkened back to Dostoyevsky and other writers who believed that we are defined by our actions, regardless of the rationalizations that might justify them. *Action* in that sense means far more than muscle movement. These authors focused on decisive moments when you put your whole self behind what you do, willing to risk everything. Such acts are fraught with meaning due to the context in which they are performed. Such acts, particularly ones involving self-sacrifice, can trigger a tidal wave of consequences. Consider, for example, Moses standing up against Pharaoh, the martyrs of the early Christian church, Sir Thomas More standing up to Henry VIII, Martin Luther rebelling against the Catholic Church, and the signers of the Declaration of Independence.

Few of us will ever perform history-changing heroic feats. But we all do affect one another through principled, heartfelt acts that serve as memorable and inspiring examples to those around us. And what, at the time, may seem an insignificant act could, through its influence on others, have major consequences.

Through our genes, we are connected to those who came before us and those who will come after us. Through ideas and chains of teaching and learning, we are connected to those who inspired us and those we inspire. And we are also connected to one another by the consequences of our meaningful acts.

The End-Game Generation Seeks Meaning

In an extended game of chess, there comes a moment when the tree of branching possibilities reverses — you visualize the ideal end position, and instead of planning ahead, you begin to plan backwards, figuring out how to get to that ideal end position.

As those of us in the baby boomer generation retire and start new lives, we find ourselves in a unique position. Previous generations knew that they would die, but, for the most part, had little or no idea of how or when. Advances in medicine and genetics are now making possible early diagnosis of long-term fatal illnesses. New treatments can postpone the onset of such conditions and slow their progress. But it will probably take decades before cures and effective preventive measures are developed. That means that many of our generation who learn that they have a long-term illness will live for years with that knowledge, with everyday reminders of impending disability and death.

How will we meet the challenge of knowing that we are in end game? Will that knowledge change how we choose to live the rest of our lives and how we perceive the meaning of our lives?

I suspect that we can learn something of value from this experience and pass that on to future generations who may not be subject to such illnesses. Then the stories of our lives might provide insights into human relationships and into how we should live and who we can become.

Truth and Consequences

Over time, perspectives change. What we are proud of today, one day we may regret. And what we regret today may one day make us proud.

In many cases, if we knew beforehand the long-term effects of what we were about to do, we wouldn't do it. But further in the future, the effects could be the reverse, and what we now dread might in a new context seem good and necessary.

As Heraclitus observed, you can never cross the same river twice. If you could relive any moment of your life, it wouldn't be the same moment, because your knowledge and your motivation would be different.

When my Dad was 86, he had trouble sleeping. In his dreams, he revisited the decision points in his life and wondered why the consequences of his decisions turned out one way rather than another. He wondered whether he had made the right choices, and what could have happened if he had acted otherwise. He was heavy with regret.

I shared with him my belief that we all have natural proclivities, and that what seem like decisions often aren't decisions at all. In our gut, we know what we have to do because we are who we are. The reasons we give for our actions are often rationalizations we cobble together afterwards. Yes, random events affect our lives. But, in many cases, such events only knock us off track temporarily, and then we continue toward the same goal by a different path.

There's a shape to the landscape in which we live our lives, with mountains and valleys. As we approach a decision-point, if we go in one direction, everything gets more difficult and painful — we trip over ourselves; we can't find the words; we forget things that we have to remember; we are at odds with ourselves. And in another direction the path feels right. If we go the first way despite the obstacles, we'll face another choice and another. And sooner or later we end up doing what is natural for us.

Hence, we shouldn't judge what we do based on what we believe will be the long-term consequences. Rather, we should do what we feel is best for now and do it to the best of our abilities.

Our lives are not as subject to random occurrences as at first appears, nor are we as much in control as at first appears. There is more to our lives than we are ever likely to realize, and that should inspire wonder, curiosity, and reverence.

Scared to Life

One night, I saw three hoodlums with machetes walk through the outside wall of my second-floor bedroom. I thought I was awake. I screamed.

It took a while for me to realize it was a vision, and for my breathing and heart rate to slow down. In the process, it occurred to me that I had come close to being scared to death. Then I realized that I had been *scared to life*.

A dream like that — not an ordinary dream composed of images from everyday life, and not a recurring dream heavy with symbolism, but one that comes out of nowhere and that you see while semi-awake — must serve a purpose.

That vision was a wakeup call for me, like a near-death experience. It was a reminder of my mortality, a warning that if there was anything I wanted to do, I'd better do it. If the obvious physical signs of illness or aging aren't enough to get me going, then my unconscious will take over and scare me into life.

That dream was an affirmation of a basic belief of mine — that as individuals and as a species, self-regulating mechanisms come into play, pushing us toward balance and reason and compassion.

Trust Your Instinct

A friend of mine was faced with this puzzle in an Ethics class:

- You are in a speeding trolley.
- The trolley can't stop.
- Five people are standing on one track, and one person is standing on another.
- All you can do is switch to a different track.
- If you do nothing, five people die.
- If you switch tracks, one person dies.
- What do you do?

It would be a different moral question if those at risk could influence the outcome — for instance, if one of them could opt to die so the other five might live. You could consider that all people are equal, and hence five people are worth more than one. I am uncomfortable with that approach. I believe that each life has infinite value and that you cannot add to or subtract from infinity. In addition, the real world is messy, and choices that at first seem good may lead to unintended consequences. Sacrificing the one to save the five might be the wrong decision for reasons you could not have anticipated.

You might be tempted to categorize and judge people as more or less valuable by some criterion. But that approach brings to mind Orwell's *Animal Farm*, where the slippery slope leads to such moral nonsense as "All animals are equal. But some animals are more equal than others."

You could also flip a coin or use another random method to decide. But by so doing, you would be abdicating responsibility.

Solution? You should do what spontaneously seems right. Don't overthink. Shoot from the hip. Act immediately. Feel the guilt and responsibility. Don't run from it with rationalizations and calculations. Feel the need for penance, the urge to make up for what you have done.

The decision should have consequences for the decider. You will always be in debt to those who die and those who are dear to them. You will also feel connected to those who survive, and they to you as well.

Guilt is the glue that holds society together. It should be valued, not avoided. You can never be guiltless, and you can never completely atone for guilt. Life is enriched by connections with others by love, guilt, and mutual obligation.

For survival, humans are programmed to act with a minimum of hesitation in such situations. Follow that instinct. Then accept responsibility for what you have done. Don't tell yourself that you had no choice. Those at risk all mattered to you and will still matter to you after the event. You honor them and connect to them by feeling guilt even after they are dead.

Part 2:
Identity, Memory, and Communication

Names and Naming

The Name of God has special significance in the first commandment — "Do not take the Name of the Lord Thy God in Vain," and in the Lord's Prayer — "Hallowed be Thy Name." Why this focus on the "Name of God" as opposed to God Himself or Herself?

As Kant pointed out, there is the thing itself, the unknowable essence that you presume exists outside of your mind; and there is the concept of the thing, its representation in your mind.

The human mind evolved to make practical sense of the world around us, to allow us to cope in a world that is fundamentally unknowable. We use names to organize and associate thoughts, and we relate those thoughts to our personal experience. In the beginning was the Word.

In some traditions, everyone has a *true name* which expresses that person's nature. Knowing someone's true name gives you power over that person.

Your name, whether traditionally or randomly chosen by your parents, is an empty vessel that takes on meaning over the course of your life. That name comes to stand for the unique person that you become. It also connects with others who came before who were given that same name.

The word *name* also refers to the categories which we apply to all of creation, like dog and cat, in recognition of characteristics that a set of things or creatures have in common. In *Genesis* Adam and Eve *named* all creatures.

And the word *name* can also stand for an unknowable essence — God — enabling us to talk about and contemplate what we cannot know.

The mind uses names to mirror the world. When we give names to what we encounter in the world, we set up mental equivalents that we can manipulate and compare and remember. In striving to understand these concepts, we assign meaning to them and associate them with one another, and meaning grows from what we think about them as well as from our experience in the world. With this cumulative remembered mental activity we enrich our lives and come to better cope with the experiences we encounter in the world.

By the ways we associate these concepts with one another,. we create maps in our minds that represent how we imagine the real world — not just a one-to-one association of ideas to things, but ideas of ideas of ideas — a rich tapestry of layer upon layer of associations. The names of things are far richer than the things themselves, because we can associate them with one another in our minds and we can communicate these complex ideas to others.

To name is to begin the effort of trying to understand.

By this line of reasoning, the Name of God is the first step in trying to understand what God might be.

Language: Anarchy, Dialogue, and Understanding

I dreamed that two people were hopelessly at odds with one another. Each used language differently. They understood the same words in different senses. But, eventually, through the medium of language, using words to define other words, they found common ground. Language was the medium for their reconciliation, helping bring about understanding.

When I awoke, I realized that what I had previously presumed was the weakness of language is, in fact, its strength.

The flexible, imprecise, self-referential nature of language makes the process of understanding possible. There is no absolute meaning for any word. Meaning comes from the conflict of two people, who use words in different ways, striving to communicate with one another. The imprecision of their words fosters the process of arriving at mutual understanding.

Understanding results not despite, but because of the anarchic creation of new words and new definitions of old words.

At the beginning of an attempt at serious communication, the fuzziness of the words that both parties use is obvious. That problem initiates the dialogue through which, by a recursive process, using language to define language, understanding is reached. In other words, language's fuzziness leads to clarification and agreement.

Two people from different regions or backgrounds or disciplines or genders often use the same words in different ways. Hence, when they try to communicate, they can't take the meanings of words for granted. Therefore, they need to clarify what they think and work hard to sort out what the other person means. That effort can lead to greater self-understanding as well as mutual understanding.

In other words, language is not a static tool that people use to communicate. Its ambiguities and flaws force dialogue, and only through the give and take of dialogue is it possible to communicate concepts of consequence.

Memory and Sex

(Adapted from a passage in my novel ***Beyond the Fourth Door.****)*

"Dreams are important for memory. They're part of the mechanism for translating short-term into long-term memory. Short-term memory is electrical while long-term is chemical. Often the elderly can remember events from their childhood but can't remember where they just put their keys. Short-term memory gets filled up, like a blackboard so covered with chalk that there's no way to distinguish anything new written on it. The elderly lose the ability to transfer recent information to long-term memory and hence cannot clean the short-term slate.

"That's where my theory strays from conventional knowledge. I believe that the mechanism for that transfer is triggered by sexual stimulation.

"Freud made a big deal about the connection between dreams and sexual fantasy. He presumed that sex is the be-all and end-all, at the center of our creative activity. I put the emphasis on memory instead. We dream not to fulfill unconscious sexual desires, but rather to renew our memory so we can continue to function as productive human beings. There's a connection between erotic dreams and the mechanism of memory translation. We dream about sex not because sex is important, but because memory is important. Sexual arousal triggers a set of events that puts our short-term memories into long-term form and then erases the short-term slate so it's ready to record new experiences.

"Other mammals have brief periods when they are in heat and can conceive. Adult humans are perpetually subject to sexual stimulation and sexual dreams, even after they can no longer procreate. That's because sexual stimulation is part of the mechanism for renewing memory. And memory, rather than sex, is the primary human drive. My hypothesis could lead to research that might alleviate or cure memory problems of the elderly."

Random Thoughts About Thought

Thought exists before words. When we associate a thought with words, the thought becomes clearer through the association of those words with other words and other thoughts. And the thought becomes richer when we share those words with other people, who then associate them with other words and other thoughts and engage in dialogue with us and with one another.

What once was a shapeless glimmer of a thought sometimes spreads from person to person and from generation to generation, gaining momentum and following its own trajectory, far beyond its origins.

When I was young, I believed that until it was expressed in words, a thought was not yet a thought, as in the opening of the Gospel of John, "In the beginning was the Word."

But as I get older, I sometimes grope for words to express a thought after that thought has taken shape in my mind. My first choice of words might be *on the tip of my tongue*, but out of reach; and I scramble to find alternate ways to express what I mean.

So, with advancing age, it's not just retrieving memories that becomes difficult, but finding words as well. The thoughts are there, but the words are not.

Perhaps that passage from John would be better translated, "Before the beginning was the Thought, and the Thought was with God, and the Thought was God…"

Anti-Reason

In the opening sentence of the *Gospel of John*, the standard English translation reads "In the beginning was the Word and the Word was with God and the Word was God."

In that oft-quoted passage, *Word* is from the Greek *Logos*, which can also be interpreted as logic or thought or reason. And *with* is from the Greek *pros*. In classical Greek, *pros* plus the accusative means *against.*

Hence, I am intrigued by this other possible translation: "In the beginning was Reason and Reason was against God and Reason was God." In other words, Reason by its very nature is opposed to God but, in fact, is God.

Coping With a Brain That Changes Over Time

We evolved to cope with this world — not to understand it, but rather to deal practically with what we encounter. There is no reason to presume that we have the capability to understand this world, much less distant worlds. And over time, our apparatus for thinking becomes less able to deal with new information and new circumstances, not just because we are older, but also because what we learn shapes how we think and what we perceive in the future. Learning and experience change us.

The typical challenge to Descartes' assertion "I think therefore I am" is that the statement presumes the existence of a subject who can think. It also presumes that the subject, the pronoun *I* — whoever or whatever does the thinking — remains static. And that is not the case. Since our thinking apparatus changes over time, if I define who I am by how and what I think, then I am a different person today than I was 50 years ago.

As William James observed in *Talks for Teachers About Psychology*, if and when we focus on one kind of activity or one realm of knowledge, it becomes increasingly harder for us to learn new subjects that are unrelated to that. We might postpone learning calculus or a musical instrument or a foreign language, only to discover later that we no longer can make sense of or enjoy or master that new subject.

I wonder if I would have made other life choices if I had understood that principle when I was in my teens, 20s, and 30s. As it is, I'll never master Japanese or the clarinet or make sense of advanced math and science.

A pessimist would say that our ability to think deteriorates as we age, as does our ability to see and to hear. But my instinct tells me that what matters overall isn't the individual mind, but rather the results of our collective thinking and what we do based on that thinking: that the aging mind has characteristics that are important in interaction with other ways of thinking. I believe the world needs older thinkers, whose minds have been shaped by a variety of learning and experience, just as it needs young thinkers whose minds are more malleable.

The Nostalgia of Tomorrowland

On my first trip to Disney World back in 1978, Tomorrowland struck me as dated — embodying an obsolete image of the future, the future we imagined in the1950s. This was the future that Disney and General Electric once promised us. "Progress is our most important product." "Live better electrically." Back then, on television, we heard about the future products of inevitable progress. Technology was marching steadily forward. Machines were making better machines to make better machines. Man was the passive spectator and beneficiary of inevitable progress.

In 1978, I expected to see an updated image of tomorrow in Tomorrowland. Surely, the people who built it intended this land to represent the tomorrow of the present, not the tomorrow of the past. But this Tomorrowland was a duplicate of the 1950s Tomorrowland. It was yesterday's tomorrow.

Then I was struck by nostalgia for the 1950s, for a time when we could believe:

- in ever-expanding resources , energy wealth, and progress;
- sooner or later (perhaps in our lifetime) there would be regular passenger flights to Mars and beyond;
- costs would inevitably to down due to increasingly plentiful energy; and
- time-saving products would be available at lost cost, as one innovation led to another.

Instead, I was living in a time when:

- costs, and especially energy costs were soaring;

- industrialization and the use of gas-powered vehicles generated pollution; and
- exploration of outer space was too expensive to be feasible;

We had to cut back and slow down. We had to abandon many time-saving conveniences that we'd grown used to as we strove to reduce our energy and resource consumption. A generation that was promised inevitable progress was forced to retreat before the consequences. The quest for progress led to the unnecessary destruction of important resources, but we couldn't help but feel nostalgia for those halcyon days when there were no clouds on the horizon, and it was all-systems-go.

Now, in 2022, thinking back to that visit over forty years ago, I remember the huge artificial tree in Adventureland, representing the home of the Swiss Family Robinson, and that memory diverts my speculation about the future in a different direction. That display showed examples of nineteenth-century ingenuity working with, taming, and living in harmony with nature. Ironically, that celebration of natural living was centered on an artificial tree.

Now, that treehouse calls to mind the ingenious methods that people in the past used before they had access to electrical machinery and internal combustion engines. I'm amazed at what they could accomplish with so little.

We can no longer afford the luxury of passive consumption. More and more, each of us must struggle to cope with decreasing energy supplies and increasing costs. We need to make the most of the objects around us. We need to recycle, turn out unneeded lights, insulate the attic, patch and fix clothes and gadgets that a few years back we would have replaced because replacement cost less than repair.

I see an end to *future shock* coming with the end of passive progress. To thrive now and in the future, we need to become handy, persistent, patient, and ingenious. We need to develop traits and abilities and learn everyday skills that our ancestors took for granted.

A new picture of the future emerges — a positive and active future I can look forward to, identify with, and participate in.

Why Bother to Save Halloween?

(I wrote this essay in 1984, when my son Bob was nine years old. Bob now has children of his own, and the message still rings true.)

Halloween is in trouble. Each year editorials in magazines and newspapers and on television warn of dangers to children. And each year more communities ban Halloween.

So what? Who needs it? What is Halloween anyway? It's just an excuse for big kids to make trouble, little kids to eat too much candy, and candy companies to peddle their wares. Bah, goblin-bug!

Or so I thought until, despite all the warnings, I took my three children out last Halloween. Nine-year-old Bobby was the boldest. Seven-year-old Heather held back and was reluctant to approach houses of near neighbors she didn't know; but curiosity and pride in showing off her homemade Ms. Pacman costume won out in the end. She went racing after her big brother and was just as delighted as he was at the smiles and words of praise and handfuls of candy that greeted them. Three-year-old Mikey held me tight and wouldn't let me put him down, but he wouldn't let me take him home either, watching all the doings intently.

As in previous years, many of the people we visited we only saw at Halloween, even though they lived a few doors away. Most of them, the elderly especially, had bought supplies of candy and were waiting. Most gave out two or three times as much per kid as they had originally intended, because there were so few kids trick-or-treating. And they were as delighted to see the kids as the kids were to see them.

When I got home, I couldn't help but wonder what had gone wrong with Halloween. And it occurred to me that it wasn't just a handful of crazy people who were endangering this tradition and the joy it can bring to children and adults as well. It's apathy on the part of those who have failed to recognize the important role Halloween plays in our society and are unwilling to defend Halloween when the media were so unanimously against it.

So, what's so important about Halloween? Maybe, at one time, Halloween helped exorcise fears of death and ghosts and goblins by making fun of them. Maybe, too, in a time of rigidly prescribed social behavior, Halloween was an occasion for socially condoned mischief — a time for misrule and letting loose. Although such elements still remain, the emphasis has shifted, and the importance of the day and its rituals has grown.

Nowadays, people often don't know their near neighbors, much less the neighbors a few blocks away. For little children, these strange houses and strange people are a source of fear and anxiety. Children have been taught not to trust or talk to strangers, to beware of them. But on Halloween that prohibition is lifted; and, fearful, but impelled by curiosity and greed for candy and other loot, little ones ring doorbells at houses of strangers to find time and again that these strangers are friendly, like the people they know. In the course of the evening, they gain confidence in themselves and in their neighborhood and come away not only with bags of candy to be enjoyed for weeks, but also a warm feeling about their neighborhood and people in general.

As for adults, especially the elderly and those who never had children or who haven't had young children at home for some time, children in the neighborhood are normally a source of anxiety and distrust. What mischief and vandalism might this new generation be capable of? On Halloween night, their fears, too, are exorcised, as wildly and imaginatively costumed kids parade to the door, reminding them of what they themselves did as children — a common link of experience.

Looked at another way, Halloween, by a ritual act of trade, reaffirms the social bond of a neighborhood, particularly the bond between strangers of different generations. Children go to lengths to dress up and overcome their fear of strangers in exchange for candy. And adults buy the candy and overcome their distrust of strange children in exchange for the pleasure of seeing their wild outfits and vicariously reliving their own childhood adventures.

In other words, the true value and importance of Halloween comes not from parading in costumes in front of close friends and family, but from this interchange with strangers, kids building a social bond with people in the neighborhood who they don't see very often.

So, when you hear warnings about pins and poison, use caution and common sense. But don’t abandon a tradition that you yourself loved as a child, that your own children look forward to, and that helps preserve our sense of fellowship and community with our neighbors amidst this madness.

Getting Lost in London or Education by Misdirection

Our hotel was just three blocks away, but I made a wrong turn and was lost for hours, walking in circles. The streets changed names every few blocks, and the intersections were rarely square, often curved or bent. I had my cellphone with me and could have checked Google Maps. I also could have asked for help. Instead, I enjoyed the scene and thoroughly learned the neighborhood.

It occurred to me that here was an excellent teaching method — Education by Misdirection. Make the path to learning difficult rather than easy. Force students to figure things out on their own or in collaboration with one another.

When my son, Bob, a chess prodigy, had lessons with a local master, Murray would often make outrageous moves and challenge Bob to find ways to punish such mistakes.

And my fifth-grade teacher, Robert Talbott Stevenson, deliberately included an ambiguous question in a hundred-question multiple-choice exam. "There are (many, few) gases in the atmosphere." I knew that the atmosphere consists mainly of nitrogen and oxygen, but there are also trace amounts of lots of other elements. So, of course, I chose *many*. I was shocked when he marked that wrong. I challenged that decision, but he wouldn't change his mind.

For several weeks, every day after school I consulted encyclopedias and other reference works, gathering evidence to make my case. And the next morning, I would present my findings at show-and-tell, but to no avail. In my outrage at the teacher, I got so caught up in the question that I nearly decided to become a scientist.

Philosophic Insight from the Blind

A sightless person perceives the world differently and relates to it differently than a sighted person. The absence of sight does not mean the brain has less data to deal with, but rather it has a different mix of data — not dominated by the data of sight.

In interacting with sighted people and their physical-social environment, the blind develop correspondences between what they perceive and what sighted people describe. Hence, the categories by which the blind organize and deal with the world approximate the categories of the sighted.

But the categories themselves are not essential features of human thought. Rather, they are learned. And the philosophic problems that arise from such categories — similarity, size, shape, color, beauty, time, etc.— are contingent on my having eyesight and having developed the associated mental processes for dealing with that type of data.

A newborn baby does not have sight-related categories. Over time, through the practice of dealing with sight-dominated perceptual data in a sight-oriented world, a young child learns to understand the world in terms of size, time, similarity, etc. Hence, we believe that learning progresses in stages, as children develop the ability to think that way. We learn how to pose questions, how to answer questions, what to accept as proof, and what to believe without question.

In other words, our fundamental concepts of space, time, and self are not structures of our minds when we are born. Rather they are contingent, learned, and alterable.

In different circumstances, without the same faculties of perception or confronted with different kinds of perceptual data, perhaps on a different planet, the human mind could develop in different ways, with different kinds of understanding, different certainties, different questions, different answers.

How I think depends on how I perceive as well as on what I perceive. How I think also depends on how I have learned to organize and process what I perceive.

In the absence of a primary sense such as sight or hearing, other senses and intuition come to play more prominent roles, leading to capabilities outside the range of the basic five senses, and leading to structures of processing and understanding outside the range of the sighted.

The world is much richer than we normally presume, and our ability to perceive and understand extends over a wider range than philosophers have presumed. We should turn to the blind for such insights, and, perhaps, in the future, we might also benefit from the insights of computer-based entities which operate with very different modes of perception.

William James and Laura Bridgman

In an article in the New Yorker, July 2, 2001, Louis Menand wrote about Laura Bridgman, a predecessor of Helen Keller. Fifty years before Helen, Laura (1829-1889) was the first blind-deaf person to get an education in English.

Her mind had very limited content to work with — mainly memories of what she could touch with her fingers. "Yet she found life as intensely absorbing as anyone else does." From her example, William James concluded "that the relations among things are far more interesting and important than the things themselves." He continued, "All sorts of terms can transport the mind with equal delight, provided they be woven into equally massive and far-reaching schemes and systems of relationship. The schemes and the systems are what the mind finds interesting."

James believed that there are many "realities" in the universe, and "that we sense relations as much as we sense things." According to Menand, Bridgman insisted that she had a "sense of think." For her, thinking was "as immediate and spontaneous as sight or touch." He concluded, "[Thinking] is the way we weave the sensuous tapestry of the world. From a cosmic point of view, all minds are pathetically underpopulated. We somehow intuit a world from a tiny sample of what is out there — not as tiny as Laura Bridgman's but possibly not as great as we would like to imagine, either."

The Decoy and the Scarecrow

Hunters use decoys to lure ducks to the target area, by making them think that others of their kind are safe there. Farmers use scarecrows to send the opposite signal — that humans are near; stay away. These are examples of rudimentary communication between humans and animals, signals of inclusion or exclusion that can be used to deceive. The same principles may be at the root of human communication.

The underpinnings of two-way communication — the ability to recognize differences of facial expression, tone of voice, and gestures — make possible the training and domestication of dogs. A dog teaches its owner which looks, tones, gestures words will most readily get a response, and what rewards and punishments will be most effective. The sensitive, experienced trainer recognizes those signals and adjusts his or her approach accordingly.

That observation makes me curious about the development of intimacy between caregivers and infants as establishing the basis for communication — setting up unique interpersonal signals, tones, words, gestures, subtle differences in facial expression and eye contact that evolve into modes of communication. These signals might only be understood by the particular individuals interacting with one another. And the use of such signals helps identify friend or foe and how close the friend is or how inimical the foe.

Establishing rapport with an infant and using non-verbal modes of communication serves as the basis for human learning. Some learning disabilities can probably be traced to early failure in the caregiver and infant connecting with one another. A caregiver who is often absent or drunk or on drugs or is apathetic, unresponsive or malignant could disrupt trust and intimacy, which is the basis on which learning is built.

Also, spoken language has layered effects. To speak the same language, to speak it in the same dialect with the same accent and the same modulation of tone establishes the friend or foe scale and says far more than the lexical meaning of the words. The speaker, consciously or unconsciously, uses codes whereby gestures, looks, expressions and tones can convey that what is meant is different from or even opposite to the meaning of the words. Layered on the lexical meaning of words, such signals indicate the degree of trust between speakers. Their use of language establishes their relationship to one another as one of inclusion or exclusion and the degree of distance.

Language might first have been different from one family, clan, tribe, nation to another. Its use established zones of inclusion/exclusion and friend/foe. The adoption of universal languages like English and the gradual homogenization of accent and disappearance of dialect leads to the creation of alternative modes of exclusion/inclusion — sub-languages, jargons, graffiti, abbreviations, and symbols that only an in-group understands.

The way we speak sends signals to one another like the signals of decoy and scarecrow. In other words, the way we communicate excludes or includes deliberately, and can be the basis for deceit as well as mutual understanding.

It's Time to Remember

My friend lamented that she wouldn’t be able to remember all that she needed and wanted as she got older. I laughed. "That’s easy. Just build new associations. Cluster together similar memories and build new hierarchies. The categories and facts that you learn in school grow faint with age. You should refresh your memory with new associations you discover for yourself.

"For example, when in a museum, focus on the artwork rather than on the printed descriptions and seek your own associations — similarities and differences, trends, styles. I recently went to a Munch exhibit in Paris and then to the Louvre. I was struck by Munch paintings that evoked emotion rather than depicting reality as perceived. The descriptions labelled them as *Expressionist*, a term that meant little to me. Then at the Louvre, I realized that medieval icons of Christ and saints had a similar purpose — evoking emotion, as an aid to meditation. Now whenever I see the one kind of painting, I’ll think of the other and remember and enjoy them all vividly."

Memory Clusters: Trying to Rescue the Past

I sometimes have trouble remembering names that I'm familiar with — friends I haven't seen for years, authors, political figures, names of movies and of actors and actresses.

This isn't the kind of forgetting that Freud describes in *The Psychopathology of Everyday Life*. That kind of forgetting derives from an association between something we want to forget and the name we now want to remember. The repressed memory accidentally obscures other memories. The association could be sound, shape, or some other characteristic. With that kind of forgetting, we often can identify the repressed memory by free association and then remember what we want to remember. Problem solved.

But in this case, even if and when I do remember the name, I don't know why I forgot it; and I might forget it again tomorrow. To deal with this kind of issue, I have started what I call *memory cluster* lists.

Often, when I forget a name, I know the context. I know that I met him or her when I worked at DEC or when I lived in New Hampshire or in association with friends we had in common. My *memory cluster* file includes the names of people I remember, organized by what these people have in common or the era of my life when I encountered them. When I can't remember, I scroll down the list to the relevant category, and when I see the name, I know that's the name I'm looking for. And the act of adding to the list or referring to it often leads me to remember other names that I hadn't recalled in many years. The more names in a given cluster, the sharper the memories and the more additional names occur to me.

Part 3:
Understanding the World We Live In

In the Beginning Was the Need to Eat and Drink

After his stroke, my Dad couldn't speak. At first, the brain damage didn't affect his ability to swallow. But after four years of not speaking, the muscles of his throat weakened and atrophied, and there was nerve deterioration as well. He could no longer swallow.

That observation led me to wonder if speech might be a secondary rather than a primary evolutionary development.

Making noises with the throat may exercise muscles needed for swallowing. It's possible that the ability and desire to make sounds through the throat evolved because strong throat muscles are necessary for eating and drinking. And the ability to generate a wide range of sounds also made it possible to use sound for communication, further improving survivability.

In the beginning was the need to eat and drink. The Word came later.

Evolutionary Quirks

Consider the evolutionary value of the beer-belly gut. Your metabolism changes as you get older, such that you tend to get fatter even if you eat less and exercise more than you did before. That makes you less physically attractive, and hence less likely to stray from your mate. In other words, your physical deterioration makes it easier for you to shift your focus from seeking new sexual partners to caring for your family, with obvious survival benefits for your off-spring.

Also, consider the survival value of physical discomfort. Your arms are short, and their flexibility is limited, in contrast to other primates. You can't scratch your own back. You need someone to scratch it for you. Depending on others to help you builds social relationships which are important for survival.

And, of course, human babies are born helpless. The more helpless the better, because that builds a stronger parent-child bond.

Bird Color

Walking to the beach today, I saw a pair of mallard ducks, and I was struck by the obvious contradiction of high-school-level evolution lessons.

The coloring of the female is dull brown, as camouflage, so she won't be seen by predators. And the male is bright green to catch the eye of females, for mating purposes.

These opposite strategies, by birds of the same species, are both supposed to promote survival of the species.

The Future of War

In the past, war had evolutionary benefits, spreading both genes and ideas. But thanks to advances in transportation and communications, ideas and genes now spread and mix globally without war.

War forced people to migrate either as attackers or victims. Without war, people might have stayed put from one generation to the next, with little contact even between neighboring villages. War led to inter-breeding as opposed to in-breeding, reducing susceptibility to illness and genetic defects due to commonality of genes. And, without the frantic competition of populations at war, there would have been less incentive for technological development. In other words, war resulted in benefits for mankind, like the benefits of cross-pollination for plants.

Mankind evolved when war was necessary and although war no longer benefits human development, it could be many years before our impulse for it atrophies. Meanwhile, wars may become less frequent, less intense, and less costly, as the urge to war no longer has a basis in evolutionary necessity.

Is Entropy an Illusion?

According to the Second Law of Thermodynamics, any closed system, including the universe, perpetually moves toward greater disorder. But I battle disorder every day — organizing books, straightening my apartment, organizing my physical world. So I question the credibility of that law.

Perhaps order is in the eye of the beholder.

According to chaos theory, complex and apparently random physical phenomena, such as weather and ocean currents, exhibit patterns that computer analysis can reveal. And a teenager may see order in what, to his parents, appears as the chaos of a messy bedroom.

Perhaps the universe as a whole, seen from a different perspective, has a more complex and not yet recognized order. The universe may have an inherent tendency toward greater order, counterbalancing the increasing disorder that scientists now observe.

The Niche Theory of Evolution

In a given territory, there are a finite number of niches in which living creatures can thrive. Species compete and evolve in those niches. Once a species dominates its particular niche, no other life form will have an opportunity to adapt and evolve there. Nature doesn't tend toward perfection. The first good approximation fills the niche.

Likewise, in a given culture, ideas and beliefs and their physical manifestations survive and evolve in a finite number of niches. If slaves could do manual labor, there was no need to develop machinery to do it. If the abacus was good enough for everyday calculations, there was no need to develop mathematics. When niches in a particular social structure are full, there is no opportunity for new ideas or new technology to evolve until catastrophic events break the structure into pieces. Then, over time, the pieces coalesce in new patterns, with new empty niches to be filled, often inspired by fragments remaining from the previous structure, and opening the opportunity for new creativity and technological advancement, until the new set of niches is filled.

Today, global transportation and communication are leading to the breakup of what once had been isolated and stable social structures. And the societies that fostered the evolution of the technology that made that happen are in turmoil, changing unpredictably due to tectonic shifts in economies and ways of life. Desperate masses of people with no home and no jobs or low-paying jobs and no role in society and no sense that their effort/work/creativity can provide them with a livelihood are likely to trigger a global breakup of social and economic structures.

Over time, the pieces will coalesce, and new ideas/beliefs/technology will evolve from the wreckage, inspired by remaining fragments of the previous stage, perhaps forming many separate, isolated, local structures. Or, if the infrastructure that supports global communication and transportation survives this upheaval, the new environment will be global rather than national, with a multitude of new niches to be filled. Countries may become obsolete. We may see an unprecedented surge of creativity, leading to a new stable global social structure.

Slavery and Industrialization

By the time of Archimedes, the ancient Greek/Roman world had remarkable mechanical know-how. What prevented them from developing steam, electrical, and internal combustion engines?

Perhaps slavery. If necessity is the mother of invention, where there is no necessity, invention does not take place. For whatever they needed to do, they had cheap energy readily available in the form of slaves. Hence there was no need to develop alternative forms of energy. The Industrial Revolution in England and then in New England coincided with the abolition of slavery. And in the South, industrialization could not happen until it, too, abandoned slavery.

How to Save the Bahamas and Maybe the World as Well

Hurricane Dorian devastated the Bahamas. That event brought home the consequences of global warming — rising seas and more frequent and violent hurricanes. So, should we focus our efforts on stop-gap measures to help the survivors rebuild or relocate? Or is there something we can do to make such events less likely and less severe?

I'm reminded of the movie *The Day After Tomorrow*. At the beginning, an increase of a few degrees in the ocean temperature leads to changes in the Gulf Stream and ironically to a new ice age — the opposite of global warming. That ice age is likely to last a long time because the ice will reflect sunlight, making the world even colder. My immediate thought was that they could have colored the ice, so it would absorb rather than reflect sunlight. That would cause the ice to melt.

Now, in the case of the Bahamas, we need to *lower* the temperature of the ocean. Of course, there are efforts to reduce greenhouse gases, but those require cooperation among many governments and millions of well-meaning individuals. And it would take a long time for those efforts to have a significant effect.

Might there be a way to reflect the light of the sun better than the surface of the ocean does?

Imagine thin inexpensive highly-reflective material, such as mylar, stretched across frames of hollow plastic tubing. Deploy hundreds or thousands of these units, floating on the ocean surface to the west and south of the Bahamas, each anchored by cable to the ocean floor. They can rise in response to strong wind and waves, but stay in place. The distribution of the units could be random. The aim is to reflect enough sunlight to lower the temperature of the ocean by a few degrees in the local area. Units could be added or removed to fine-tune the effect. The hope would be that the temperature differential of the water would alter the path or reduce the intensity of storms heading toward the Bahamas.

If that proved effective, a similar approach could be tried on a larger scale, to help reverse global warming.

Why Didn't God Make Little Green Mammals?

Have you ever seen or heard of a little green mammal?

Why not?

In biospheres with lots of grass and leaves, we find green reptiles and green birds, but no green mammals. If color is a survival factor for reptiles and birds, why not for mammals? Of all the species of mammals, wouldn't you expect at least one to take that niche?

It's easy to use Darwinian truisms to explain the color or structure of any creature after the fact. But what about the capabilities and the colors that, by the same logic, should exist but don't?

Mammals evolved from reptiles and many reptiles are green, presumably because of the survival value of that color in many environmental niches. It would not have taken mutations for there to be green mammals. Rather, all that was needed was for one or more species to keep their green skin color and continue benefitting from that color. Why didn't that happen? Why don't any mammals have green skin, like snakes and crocodiles and frogs?

Did Large Dinosaurs Have Two Brains?

Yesterday, I had to stop for a hook-and-ladder truck backing into a fire-station. A driver up front, working in coordination with another driver at the back, maneuvered it into a tight space. The fire truck reminded me of a huge dinosaur like a brontosaurus or tyrannosaurus rex. It dawned on me that maybe such dinosaurs were more complex and intelligent than is generally presumed. Wouldn't it make sense for such a beast to have two brains, like a fire truck?

As I recall from a recent museum visit, tyrannosaurus rex had a large bone structure between the legs that didn't seem to have a purpose, except maybe extra weight for balance. Perhaps that bone structure protected a second brain. Such a beast might have had a small brain up by the eyes, for visual pre-processing, and a second, larger one located near the middle of the body, and well-protected by bone and muscle.

This speculation reminds me of the evolution of PC design, from single processor computers to multi-processor machines, where separate chips pre-process graphics and do other functions to speed overall performance.

Why should we care how dinosaurs thought? If some dinosaurs had greater brain power than generally presumed, that would alter our notions of evolution-related *progress*. And if a two-brained species could thrive for millions or tens of millions of years, that would expand the range of what we consider possible, not just on Earth, but also elsewhere in the universe.

Alternate *War of the Worlds* Plots

Watching the Tom Cruise remake of *War of the Worlds*, alternate disaster-movie plots occurred to me.

In the H.G. Wells story, the microbes to which man has built immunity operate as a global defense system against invaders from outer space. But the same mechanism could work in reverse. When Europeans first arrived in America, they were lucky not to die of diseases that Native Americans were immune to. Rather, they brought new diseases, like chicken pox and small pox, that decimated the natives. Along those lines, a benevolent, well-meaning alien species might inadvertently bring with them microbes that wipe out the entire human race.

In another plot line, the delicate environmental balance of Earth could act as a defense. Global war against alien invaders causes devastating weather consequences that wipe out the aliens. A handful of humans survive to start afresh. Global environmental collapse turns out to be the Earth's defense mechanism.

In still another plot line, aliens might use the delicate environmental balance of Earth to their advantage, as a way to destroy mankind, without the risks to themselves that would come from war. In this story, aliens, not man, are responsible for today's global warming.

Worm

Why presume that aliens who might threaten us will be intelligent?

Imagine intergalactic worm-like creatures — long hollow tubes that swallow and assimilate matter they randomly encounter.

Imagine these creatures can grow to enormous length, but are narrow, so they are almost impossible to see by telescope.

Imagine such a creature is light sensitive. It detects targets based on reflected light.

Imagine it could eat its way through Earth.

Promised Land

Imagine there's an invasion of Earth by beings from another galaxy, and the invaders are the chosen ones of God, and Earth is their Promised Land.

Neverending Story

Nietzsche believed that after man would come superman, the culmination of evolution. But why should the process end with superman? Why not superduperman and supersuperduperman? Why sacrifice yourself for the benefit of a species that will only be another intermediate step?

Marx made a similar mistake. If history were indeed a dialectical process, why should that process ever end?

Wild Speculation — Multiple Edens and Hominin Marsupials

When I was in school, I learned that early man co-existed with and then drove to extinction a similar species, Neanderthals. Later I learned that humans (homo sapiens) evolved in Africa and migrated to what is now Europe, where they encountered Neanderthals, who had evolved separately. Later still, I learned that humans interbred with Neanderthals and that present-day humans have 1–4% Neanderthal DNA. More recently, I've seen news articles about the discovery of Denisovans and Hobbits (homo floresiensis, found on the island of Flores in Indonesia) and that they too interbred with humans.

In other words, there were several separate evolutionary paths leading to creatures close enough to humans to interbreed with them. There was more than one Eden.

Now I'm on vacation in Tasmania, reading about the prehistoric fauna here and also watching wallabies and kangaroos play in the yard of the BnB where I'm staying. Wallabies look and behave very much like rabbits in America.

Australia was once dominated by huge marsupials that filled evolutionary niches similar to those of placental mammals of Africa. It seems that, given similar physical circumstances, separate animal families evolve in similar ways

That line of thought led me to speculate that perhaps a marsupial hominid also evolved in Australia, anatomically similar to other hominids, but with a pouch for nurturing the young.

I suspect that with today's level of scientific knowledge it might be impossible to detect such a pouch from fossil evidence. But perhaps marsupial hominids interbred with Hobbits or Denisovans or homo sapiens, and hence evidence of their existence might be found in human DNA today. Hence the improbable but plausible premise for fiction that someone today might be born with a pouch.

Part 4:
Politics and Government in the U.S.

How to fix Congress? Eliminate Seniority

Congress has been deadlocked for years, accomplishing little, due to partisan rivalry. Some think term limits might change the way Washington works, but that would take a Constitutional amendment. Others look for campaign finance reform, but for every rule there's a loophole.

I suggest eliminating the seniority rules of the House and Senate. Replace those rules with another way of establishing committee memberships and chairmanships, and much of the bartering of favors would disappear, smashing personal power bases, and weakening the influence of lobbyists as well.

Members of Congress could choose which committees they wanted to serve on, and membership could be decided by lot. Chairmen, too, could be selected by lot and could stay in the post no more than a year. As a result, incumbents would no longer have an advantage over newcomers in elections, leading to shorter terms. And with no clear centers of power to focus on, lobbyists would no longer be motivated to direct vast sums of money toward particular races. By reducing the incentive for corruption, corruption would decline.

How could we get from here to there? Congress would never make such a change. The President doesn't have the authority to do it. A Constitutional amendment could bring about such a change, but that would never happen because Congress would never vote for it, and state legislatures, which have to ratify amendments, have the same kind of seniority rules, with similar entrenched power structures.

But there is a practical solution. Today, with seniority rules, some elected representatives have far more power than others. If my district has a freshman congressman, I am not fully and equally represented. Hence a group of citizens could bring a class-action suit against Congress, challenging seniority rules. Then the Supreme Court could decide the issue, on the basis of the principle of one person one vote.

How to Fix Congress? Limit the Powers of the Leaders

Contrary to the intent of the authors of the Constitution, power has been shifting from our elected representatives to individuals in leadership positions. By Congressional rules and practices (not the Constitution), both the Speaker of the House and the Majority Leader of the Senate now wield extraordinary power. In particular, they decide which bills are debated and which are voted on. In the case of the Senate, the Majority Leader alone can prevent any nomination to the courts or the cabinet from ever coming to a vote. In the case of the House, the Speaker can prevent any appropriation bill from ever coming to a vote.

In the past, the two parties sometimes worked together, civilly, to arrive at compromises satisfactory to both. But in the current partisan atmosphere, the party leadership stands in the way of individual members of the House and Senate exercising their rights and responsibilities by voting on legislation and nominations based on their knowledge and convictions as well as the interests of their constituents.

Several inequities need to be dealt with promptly:

1) When a bill is passed by one house of Congress, it should be debated and voted upon, promptly, in the other house.
2) When a bill is entered in either house with the endorsement of at least a third of the membership of that house, said bill should be debated and voted on promptly.

3) When the President makes a nomination which requires confirmation by the Senate, that nominee should go through hearings and be voted upon promptly.
4) When the President introduces legislation, both houses of Congress should debate and vote upon it.

The Speaker and the Majority Leader should not be able to block such votes and related debate.

Limiting the power of the Speaker and the Majority Leader would restore power to individual members of the House and Senate, giving them the opportunity to vote and to be heard, rather than being reduced to the role of mere tokens in partisan battles. And increasing the power of individual members would enable the houses of Congress to fulfill their constitutionally mandated role of overseeing and checking the actions of the Executive.

Given the present power structure, it would be impossible for Congress itself to make such changes. It would never come to a vote. But as this problem arises from abuse and perversion of the Constitution, the courts should be able to remedy it. A court challenge should focus on the rights and responsibilities of individual members of Congress, restoring to them the ability to stand up and be counted. The current rules and practices disenfranchise them. This in turn disenfranchises the voters who support them.

Long-overdue Constitutional Amendment

(written 11/28/2019)

Many people presume that the most memorable sentence from the Declaration of Independence is embodied in the Constitution. It is not, but it should be.

"We hold these truths to be self-evident, that all men are created equal, that they are endowed by their Creator with certain unalienable Rights, that among these are Life, Liberty and the pursuit of Happiness."

The founding fathers could not agree on including that sentence in the Constitution. To make up for that, they later passed the Bill of Rights, as amendments. But those amendments were far more limited than that one bold statement in the Declaration of Independence. Can we do any better today?

The concepts expressed in that sentence could help guide Supreme Court decisions on important issues. The words would need to be fine-tuned for clarification and to make them consistent with current beliefs and parlance, for instance —

- There is no need to say, "We hold these truths to be self-evident."
- Subtitute "all citizens" for "all men."
- Substitute "are equal under the law" rather than "created," to avoid the religion issue and to clarify the nature of that equality.
- Delete "by their Creator" to avoid the religion issue.

The sentence would then read: "All people are equal under the law. They are endowed with certain unalienable rights, and among them are life, liberty and the pursuit of happiness."

This sweeping statement would need further clarification. For example,

- modify the right to liberty to allow for punishment for crimes;
- stipulate that one person's pursuit of happiness should not impinge on the rights of others; and
- define who is a citizen entitled to these inalienable rights.

In this contentious, polarized time, can we find a way to embed that basic statement of rights in our Constitution, for the benefit and protection of future generations?

The Role of Vice President

If the president dies or is removed from office or is incapacitated, the vice president becomes president. Otherwise, the vice president has very little to do. He/she can preside over the Senate and cast the deciding vote in case of a tie but has no other official duty.

Frequently, former vice presidents run for president. But, typically, while serving as vice president, they do not get the kind of experience that would make them good candidates for president. And former vice presidents have rarely won presidential elections.

The post of vice president often seems like forced retirement. A capable and well-known politician is chosen as a running mate to help win some state or block of voters. Then he or she is in limbo until the next election.

In recent history there was one interesting exception. Because of the personal relationship between Dick Cheney and George Bush, Cheney played a major role in decision-making. While I did not agree with many of the decisions that duo made, I found their division of labor intriguing and promising.

Because of his age, Cheney was not aiming to run for president himself. And because of his previous experience, he had greater understanding of the workings of government and the complexities of foreign affairs than did Bush. The fact that they shared power had nothing to do with the Constitution. It happened solely at the will of Bush.

In today's world, the job of president is enormous, far too complex for any individual. The best presidents choose excellent advisers and cabinet members and delegate authority rather than depending on their own knowledge and judgment. Bush seems to have elevated this style to a new level, sharing the power and the burdens of the office with his vice president, apparently without the rivalry that often arises from the sharing of power. And if anything had happened to Bush, Cheney would have been ready to govern on his own, immediately, without an extended learning period, which could be dangerous in time of crisis.

I hope that future candidates for president will be wise enough, confident enough, and humble enough to choose more experienced running mates who can serve as mentors, and whom they can rely on to share the presidency with them from day one, rather than simply waiting in the wings.

Delay, Rumination, and Compromise in Government

In the US, the process of making laws is long and complicated. It typically involves public hearings, committee meetings, open debate in the House and Senate, passage by both chambers, the signature of the President, and, in case of legal challenge, approval by federal courts, up to the Supreme Court.

If a piece of legislation is important to you, the issue seems clear, and the process feels broken. You want immediate action, and instead it takes months or years before anything happens, and the resultant law is often a compromise, rather than the measure that you were hoping for.

Originally, the electors who voted for president and vice president were selected by state legislatures, rather than directly by the citizenry. That was later modified so the selection of electors was determined by popular vote. Perhaps now legislation should also be approved by popular vote.

With the universal availability of the Internet and smart phones, it is possible to bypass representative government and, at least in some cases, substitute direct referendum, as a clearer reflection of the will of the people, as true democracy.

But referenda appeal to the emotional mob response of the uninformed public. And with referenda, only one specific version of a measure is presented for vote, without the opportunity for amendments or compromise. That process puts the true legislative power in the hands of those who write the specific words that the public votes on, which makes referenda a popular tool of dictators seeking to create the illusion of public support and democratic process.

While the Constitution was written in a different era when communication was slow and direct referenda were impossible, the principle of representative government is still important. Theoretically, the public elects individuals whose judgment they trust, and those representatives strive to understand complex issues and arrive at agreement with representatives from other regions and constituencies. The lengthy multi-staged process provides opportunities for all opinions to be expressed and weighed and for the wording to be fine-tuned many times. Delay forces rumination and compromise, which are essential for democracy.

A PRISM Solution

(I originally wrote this on November 15, 2014, as an open letter addressed to Elizabeth Warren. Now I'm binge-watching Newsroom *and finding that show eerily relevant to today's political nightmares. It has also motivated me to be a* Greater Fool, *like McEvoy in the series. Hence, I'm resurrecting this idea in hopes that now it might get traction and lead to important changes. If you like this idea, please spread the word.)*

Dear Senator,

Please consider filing a lawsuit under the Freedom of Information Act for release to you of any information about you that has been collected and stored as part of the PRISM program.

The entire database is a matter of national security, but your small piece of it is not. The information you would be seeking is for your use alone. Moreover, release of that information to you is not a breach of your privacy.

You should file such a law suit:

1) to determine what is in fact being collected, so you can make appropriate judgments in your oversight role as senator; and
2) to set a precedent for personal access to PRISM data.

At this point it is impossible to do away with PRISM. But the data so gathered might prove useful to innocent people whose privacy rights have been violated.

Individuals should periodically be given access their own information:

1) as a way to verify that the information is correct, rather than a result of mistaken identity or misinterpretation (comparable to access to one's own credit reports);
2) as a tool in uncovering and addressing instances of identity theft;
3) as a source of evidence for defendants to establish facts and state of mind in criminal and civil trials;
4) as a source of raw data for input in personal profile software, so individuals could build a predictive model of themselves in order to better understand their own preferences and behavior patterns, get a better idea of who they are, and have better control over their own lives.

If individuals had that kind of access to their own data, they could come to see such data collection as having positive benefits that, in part, counterbalance the loss of privacy.

P.S. It is high time that Snowden be pardoned, compensated for the misery he been subject to, and awarded a medal of honor.)

Rolling Boycott — a blueprint for turning public outrage into effective action

(I'm binging on ***Newsroom*** *for the third time now. I just got to the episodes about Occupy Wall Street. That reminded me of something I wrote back on Oct. 17, 2014, during democracy protests in Hong Kong, a blueprint for how ordinary citizens, working in unison, can change the world.)*

Occupy Wall Street in 2011 and the Hong Kong democracy demonstrations in 2014 each captured the attention of the media and won the sympathy of millions of people, but then fizzled out with no tangible results. They could and should have given their well-wishers ways to help their cause.

In the 1950s and 1960s, the Civil Rights movement used boycotts as a way to turn public outrage into effective action. But in the pre-Internet era, that process took decades. We can't afford to take decades to deal with global warming. And ease of access to the Internet and its social media platforms makes it possible to quickly implement boycotts on a global scale, and to make them *rolling boycotts.*

Focus on one clearly-stated issue central to the global warming crisis and mobilize not just to get publicity, but to force change.

Pick one well-known company whose practices you believe must be changed.

Declare a worldwide boycott on that company's products.

If after two weeks that company does not take the action you require, call for a sell-off of that company's stock and expand the boycott to include one or more of that company's largest business partners.

If there has been no progress two weeks later, expand the boycott to a second company, with the same pattern of escalation.

Continue adding companies to the boycott until your demand is met.

With that approach, you give your well-wishers a way to help you. That can be far more effective than mass marches or sit-ins or advertising campaigns.

May the Force be with you.

Tax Credits for National Service?

We need to encourage volunteerism. There are many kinds of work that need to be done but don't get done because government or charities can't afford to pay for them. And there are many people who are unemployed who could help and would want to help if there were some tangible benefit for doing so.

Redefine *national service* to include volunteer work for charities. The more national service you do, the more you receive in tax credits, and you can choose when to exercise those credits.

For instance, if you are unemployed for six months and do volunteer work during that time, you could cash in those credits after you find a new job and have a significant tax bill. If you do volunteer work while employed, you might want to use your credits immediately. And if you choose a national service career. such as serving in the military, you might want to save up credits for use after retirement.

With such a system in place, current charities might flourish and extend their services, and new charitable organizations might be formed to do more work that really needs to be done.

Selling Money — An Alternative to Taxes

Celebrating the British Open and the retirement of Jack Nicklaus, the Bank of Scotland issued five-pound notes with a photo of Nicklaus. Tourists, collectors, and golf fans rushed to buy them.

Also, the US Post Office (through stamps.com) let's you, for a premium price, design your own officially-approved postage stamps, for instance, with photos of your kids.

Congress should consider selling by auction the images that appear on US currency, as an alternative way to raise revenue and reduce taxes.

They could start with pennies. It now costs one and half cents to mint a penny. In other words, the government loses money making pennies, and it's likely that pennies will be eliminated soon. As an alternative, the government could sell advertising, putting a company logo and/or the image of a billionaire on pennies, instead of Lincoln and the Lincoln Memorial. From there, the project could be expanded to include all coins and paper currency as well.

Imagine an annual on-line auction for each denomination. Imagine all the wealthy people who might want to see their picture on the $1 bill and might be willing to pay millions for that honor.

Corporations already pay tens of millions of dollars for the naming rights to the stadiums of professional sports teams. How much more might they be willing to spend to have their corporate logo and tag line on US currency? Or imagine fans of a rock star or movie/TV star or supermodel or sports hero willing to do *class-action* bidding for the image of the personality of their choice — combining millions of small bids automatically online, like voting for baseball all-stars or for winners of American Idol.

If you could raise billions of dollars that way, every year, why not? And as a result, the currency itself could become stronger in foreign markets than it is today, simply from collectors taking millions of dollars in bills out of circulation.

And if that project works, why not put advertising on military uniforms, tanks, planes, as well as on government buildings and historic monuments. Then we would be well on our way to free-enterprise, tax-free Nirvana. Hallelujah!

Make It Easy to Make Charitable Contributions at Tax Time

As late as April 15, you can make last minute contributions to your IRA, SEP and other retirement funds. Those contributions count toward your previous year's taxes. That means you can maximize your retirement contribution and minimize your tax, using tax preparation programs like TurboTax.

Why not allow charitable contributions on the same basis? Then you could decide how much you would like to contribute with full knowledge of your tax situation, rather than guessing at the end of the calendar year. Such programs could then prompt you with suggestions of reliable charities dedicated to the kinds of activities that you want to support, and help you make the desired contributions online.

Certification for Panhandlers

When you walk out of Grand Central Station, you are immediately greeted by panhandlers with heart-wrenching stories on signs and sympathetic facial expressions. They are often men claiming to be veterans or women claiming to be abused and pregnant. On the streets of Paris, the panhandlers are usually families, including small kids, purporting to be refugees, sitting on blankets, their belongings near at hand. It's a worldwide problem.

Many passersby would give generously if they knew these stories were real. But how can you tell what's true and what's fiction? Maybe you give loose change on the spur of the moment. But there's always the lingering doubt — are you being suckered or could this be a situation like in *Slumdog Millionaire, Oliver Twist,* and *The Hunchback of Notre Dame*, where beggars are exploited, and sometimes children are deliberately mutilated to make them look more pathetic.

I'm a bit of a Dagwood, the cartoon character who frequently said *There ought to be a law.* I can't help but think of what might be done to change things for the better.

Might the omnipresence of cellphones help? Imagine a program for certifying panhandlers, perhaps run by homeless shelters, and sponsored by corporations. When the administrators are convinced that someone is truly in need, that person is assigned a QR code which he or she can display. With a cellphone app, a passerby can scan that code and see a profile of this person and what he or she needs and why, and what he or she has already collected toward that goal. Passerby could immediately make donations to the account of that particular panhandler. And sponsoring corporations could choose to match donations up to some preset limit. The money and/or credits could be collected by the panhandlers at the same shelters that do the certification.

Cultural citizenship — an alternative form of government enabled by the Internet

We have grown accustomed to thinking of the U.S. as a *melting pot,* with people from many different cultural backgrounds. But the same is true of other countries around the world. Many people feel a dissociation between their cultural identity and the political entity that rules the geographic area where they happen to live. This dissociation leads to feelings of isolation, of being out-siders, of not belonging. Where cultural minorities are large enough, that dissociation can lead to political unrest, rebellion, and civil war.

We increasingly find ourselves with two or more identities — the bonds that arise from where we are physically located and the bonds that arise from our cultural background and beliefs. Now that the Internet reaches all countries and is available to the many, rather than the elite few, it's time to consider the opportunities this communications capability opens for changing our concepts of what constitutes a government and what its role should be.

One possibility would be to recognize *cultural citizenship* in addition to *geographic citizenship.*

I was born in the United States and live here and pay taxes here and vote here. But I may feel I have a cultural identity that is German or Irish or Scottish or Italian. Someone who lives in Israel may feel cultural allegiance to Judaism or to the Arabic world.

Consider the possibilities. Say that you have the opportunity to declare your *cultural citizenship.* Then say that some part of the taxes collected from you goes to support an organizational entity devoted to your chosen culture, and you have the right to vote to determine the leadership and direction of that cultural entity. Then in a country with many cultures, no single culture would be the *winner* and all the others *losers* or *minorities*. Regardless of where you live, you would belong to the culture of your choice and could contribute toward the preservation of its traditions and help decide the direction of its future development.

This virtual, cultural citizenship need not be limited to ethnic or religious or other pre-existing cultures. People should be able to belong to whatever cultural entities they wish, including newly created ones, based on common interests. And cultural citizenship need not be all-or-nothing. Imagine you could, for example, choose to be 60% German, 20% Turkish, 10% Argentinian, 10% Mormon Church. In that case, the culture-related taxes and related voting rights would be similarly divided.

The infrastructure and the habits of behavior supported by the Internet make this bizarre notion possible. Would it be desirable? How might it work? And how might we get from here to there?

How to Deal with Dictators

(There will always be dictators, so we need to find better ways of dealing with them. The following essay was written in 1998, when Fidel Castro was still alive and still ruled Cuba, but the ideas remain relevant.)

Perhaps Major League Baseball could help put an end to tensions between the U.S. and Cuba. Baseball is extremely popular in Cuba, and Castro himself was a player with professional potential before he became a revolutionary and then dictator.

Consider the possibilities if Major League Baseball were to offer Castro personal ownership of a major league franchise in Havana. The sole condition would be that he retire from politics. This would in one swift stroke

- give Castro a way to step down, gracefully, without losing face;
- give Congress and the President an excuse to re-establish normal relations, without losing face;
- create a firm basis for quickly opening commerce between Cuba and the US, starting with business opportunities related to the games, news coverage, television rights, the stadium, travel, visitors/tourists, etc.

It could also provide common ground for understanding between the American people and the Cuban people, through their enthusiasm for baseball.

Such a move could not only reduce political tensions but also lead to establishing a Caribbean or Latin American League, having equal status with the American and National Leagues, including inter-league play and participation in the World Series.

Now consider present issues with Kim Jong Un of North Korea. What might he want? What could become the basis for negotiations for reducing the possibility of nuclear war?

Remember his fascination with basketball and Dennis Rodman. Might an NBA franchise prove tempting? Or remember the academy-award-winning move *Argo.* Might he want to star in a Hollywood blockbuster, perhaps in the role of a hero who saves the world from alien invasion?

If you take out a dictator, he gets replaced by someone else at least as nasty or you leave a power vacuum. If you pile on economic sanctions, innocent people suffer and starve. Instead, consider unorthodox solutions that play on the unique and eccentric interests and tastes of individual dictators.

How to Make the US Postal Service Competitive

Unlike its competitors (UPS and FedEx) the US Postal Service is required by law to serve the entire country, not just the most lucrative markets. It's important that everybody, including rural customers, be served on equal terms (same price and same frequency of delivery). But the cost of providing that service makes the Post Office non-competitive.

Make universal service a requirement for its competitors as well. Require them either to serve everyone, with no surcharge for rural deliveries and with equal frequency of deliveries, or to pay subsidies to the US Postal Service, proportional to the extra costs that it will incur to provide that service.

That approach would make the Post Office more financially viable, levelling the competitive playing field while ensuring equal delivery service for all citizens, regardless of where they live.

Longevity and Governance

Different kinds of creatures live for different periods of time — some for a day or less, some for hundreds of years. And while they all live in the same world, that world must look different to them depending on their average life expectancy. If you only lived for a day and on that day a hurricane hit, your priorities would differ from those of a creature with a three-hundred-year life span, for whom storms seem normal and transient.

Thanks to the foresight of our nation's founders, the US Constitution was engineered to provide a balance between short-term and long-term interests. The members of the House of Representatives serve for two years, the president for four years, senators for six, and the Supreme Court for life. But with advances in scientific knowledge, we become increasingly aware of the long-term consequences of present-day decisions. The way our governing bodies are now constituted, they cannot be expected to give sufficient weight to long-term, possibly irreversible consequences. But how can we get from where we are now to where we need to be to achieve the necessary far-sighted balance in legislation? I imagine a scifi solution.

In the socially static Middle Ages, the Three Estates — commons, church and nobility — were each represented in government. Each presented its point of view and, in the best of times, strove to arrive at constructive compromises that protected its rights and interests. In Plato's *Republic*, classes were artificially created. Citizens were trained and brain-washed from birth for the roles they were destined to play in society, under the benevolent protection of the Guardians and the wise rule of the Philosopher King. In *Brave New World* classes were genetically engineered.

I imagine a time when it will be possible to genetically engineer longevity. Then it would be physically possible for everyone to live a very long time, but that would necessitate reducing the birth rate to avoid disastrous overpopulation. Such tampering with the birth rate would make humans more vulnerable to extinction in case of some new disease or natural disaster. Instead, this technology could be used to create classes and governing bodies distinguished by their longevity.

One house of Congress could consist of people with short lives, who would champion the needs and concerns of the present and of the immediate future. Another house could consist of people who will live far longer and are more concerned about consequences that will show up decades or even centuries in the future. Perhaps there should be a third house as well, to represent the middle ground.

The Liberal Score Card

When it comes to politics, I feel frustrated, drained, lost. The news, liberal or otherwise, jumps from one issue to another. I need a sense of continuity. Don't just give me the results of opinion polls. I want to know where we stand on each major issue right now, and why, and what to expect next. I need an index of political progress in the US, based on liberal values.

First, establish goals for each issue. Assess where we stood on each in 2016, pre-Trump, and where we stand today with respect to those goals. Change the score and publicize it whenever something happens to move us closer or farther away. Assess the trend (with dotted lines) based on legislation and executive action and legal challenges that are likely to come to fruition nationally and at the state level. Make the status in 2016 equal zero, and the goal equal 100.

Cover such issues as:

- abortion/women's rights
- climate change
- voting rights
- gun control
- immigration
- racial equality
- prison reform
- fair wages
- LGBT rights

Publish the assessments of political scientists and sociologists with established reputations, perhaps under the auspices of several universities. Update the scores at least monthly, more frequently when a given issue is particularly volatile or elections or court decisions are imminent. Report the scores through a dedicated web site and in emailed newsletters, delivered free to those who want them — a separate newsletter for each issue.

Each issue starts with the score and graphs the changes over time. The newsletters are meaty, unlike the annoying emails we all receive begging for donations to this campaign or that. The concise explanations of what is happening and why and what is likely to happen next should rivet the attention of their audience. People should be motivated to forward them to friends, rather than delete them unread.

Changes in the score should become news that existing media want to report, like their coverage of established opinion polls. And the news items should serve as leads that reporters will want to follow.

Organizations that raise money for Democratic candidates could and should invest some of that money in this project. The vast majority of us tune out political advertising. Instead, show us goals that could make a difference in our lives, that could save democracy and our way of life and the planet itself. Give us scores that we can believe. Show us a path forward.

Instead of venting and hoping for miracles, let's make this happen.

Thoughts on the Overturning of Roe v. Wade

(A long-time friend of mine told me that he believes life begins at conception and that abortion is murder. He agrees with the overturning of Roe v. Wade. This is my reply to him.)

When life begins is a religious question. The Supreme Court should not favor one religion over another, should not make judgments based on religious views. Such issues should be a matter of personal choice.

Many Jews see the overturning of Roe as a violation of their religious rights. And even the Catholic Church is ambivalent on this issue, despite the Pope's stance that it is murder. The Catholic Church does not consider a fetus a human being. It only allows baptism after a baby breathes and does not allow burial in Catholic cemeteries without baptism.

The US government does not recognize the unborn as citizens, with the rights and responsibilities of citizens. A pregnant woman cannot declare an unborn child as a dependent for tax purposes. (But considering the costs involved in carrying a fetus to term, such a deduction would make sense, regardless of whether the unborn was considered a person and a citizen.)

Deciding that life begins at conception would mean that use of the day-after pill is murder. And it's a small step from that to consider birth control as murder as well as hysterectomies and even masturbation. Similarly, destruction of frozen embryos could be considered murder.

In his opinion in support of the majority, Clarence Thomas said the logic behind the overturn of Roe could and should be used to overturn cases establishing rights to contraception, same-sex consensual relations, and same-sex marriage.

The question of if and when to have an abortion should be up to the mother in consultation with doctors and nurses. Nothing in the Constitution gives the government the right to make such decisions or to curb a woman's choice in such matters. And freedom of religion, under the first Amendment, should rule out such governmental interference.

There is no need to go into all the practical matters — the instances when abortion is necessary to save the mother, the cases of incest and rape, and the fact that without access to safe legal abortion, the desperate will seek unsafe illegal abortions and many young women will die (which happened frequently before Roe).

Each case is unique, each decision personal and difficult. Let women make their own choices.

As for states making such decisions separately, some legislatures have already passed laws to make it illegal to go out-of-state for an abortion and illegal for anyone to help them do so, even making it illegal (in Florida) to offer information about such options. Laws of that kind are contrary to the First Amendment right of free speech. And the notion of setting up inspection stations at state borders to determine if women who are traveling out of state are pregnant is repugnant, as is the notion of subjecting women of childbearing age to pregnancy checks. This smells of Dread Scott and slavery.

Basing the original Roe ruling on an implied *right of privacy* was a grave mistake. Rather, the ruling could and should have been a matter of freedom of religion and separation of church and state. It should be deemed unconstitutional for any US government (federal, state, local) to decide when life begins and make laws based on that religious belief.

Part 5:
Literature, Reading, And Writing

Why I Write

One

I started writing fiction in the second grade, 68 years ago. My first story was modeled on the Kipling's *Just So Stories*. I called it *How the Parakeet Got Purple Cheeks.* By the third grade, I craved an audience and started reading my stories for show-and-tell.

Two

Stories are born, not made. Some characters and plots come to life, grow, and change. Others remain static. For an author, the experience of bringing a character to life is similar to learning a foreign language. When you immerse yourself in a language, you can reach a point when you start to dream in it. And when you develop your characters to a critical point, you come to dream what they say and what they do. You hear their words and see their actions, and the story takes on a life of its own.

Three

When I create characters and scenes, I am performing thought-experiments. If I put this set of characters together under these circumstances, what is likely to happen, and why, and how does their story shape my notions of human nature?

Four

The novels I write are memoirs of lives I have not yet lived.

Aphorisms about Reading, Writing, and Rewriting

— The first draft is the cocoon in which the story matures.

— The need to write fiction is an incurable disease you are born with.

— I write to find out what I think and believe.

— Create characters, not ideas.

— To say more, say less.

— My story ideas come from emptiness. Perhaps it's a quantum effect. When I clear my mind of everyday concerns and think about nothing, my best ideas occur to me.

— Thinking about your book is the real work. Putting it on paper is easy.

— In rewrite mode, anomalies are opportunities. Explaining them can add layers to the narrative.

— Sometimes a book happens to you, like you are pregnant with it.

— One measure of the power of an author is how little needs to happen to show the characters undergoing life-shifting changes. In the best novels, a look or a word can have the narrative power of an earthquake.

— Good writers piggyback on the attitudes, beliefs, and concerns of the day. They presume their audiences live in the same world they do. Great writers create the worlds they write about, telling enough so others — in different times and cultures — can fully imagine them.

— The aim of logic is clarity. The aim of language is nuance, ambiguity, suggestiveness. Logic seeks meaning that doesn't change over time. Language lives, grows, changes.

— The creative phase of writing is very different from the polishing and editing phase.

— To write something new or to significantly rewrite, I need to find a *generative* phrase — a line that implies a whole character, a whole life; a line that leads to another line and another and that generates a rhythm that carries the story forward. That's a very different process from analysis and criticism.

— Sometimes a good line is a hazard. You can like a line so much that you keep it, even though it wrecks the flow of the lines around it and of the story as a whole.

— The story is the vessel into which I pour my blood and guts, making exterior what's interior, so I can look at it and try to make sense of it.

— In writing, what is most private and personal is what connects us most with others, for that is what we have in common.

— The aim is to get to a state of *flow* in which what matters to you finds external expression, and that external expression triggers in others something resembling your own internal experience.

— Definition of poetry: when words explode in your mind.

— Our self-knowledge and our knowledge of others is limited. Every memoir we write is fictitious in ways we do not fathom. It is more honest to call what we write *fiction* and to shape the story the way its internal logic demands.

— If and when characters appear in your dreams, you should write down what they say and do; then edit and rewrite. Then it's their book, not yours. Treat them with respect and follow their advice.

— Once your characters come alive, you are always writing — no matter where you are and no matter what else you might be doing at the time.

— Publication does not equal success. You have to enjoy writing for its own sake. Half a million people run in marathons in the US each year. Only a couple dozen win. They simply enjoy doing it.

— I look forward to the high when characters come alive and take charge, and I'm just along for the ride.

— When you begin your novel, the characters are your means for telling the story. If and when they come alive, they become the story.

— Typos can be fun in unexpected ways. They are like random mutations, some of which win in the struggle for survival, prompting puns and stories and novels. My writing would be lifeless without the inspiration of my typos.

— To deal with memory glitches, empty your mind and let thoughts and words enter on their own. The harder you try, the harder remembering becomes. The paths, not the memories themselves, wear out. Let your mind open new paths. Control by not controlling.

— Editing your own novel is like cutting your own hair. Everything's backward in the mirror of your mind. It can be done, but it takes practice and patience.

— A novel without time travel is either short-sighted or unrealistic. Aging is time travel, and reading is as well.

— As I get older, I'm noticing a widening gap between what I intend to do and what I do. I think about what to do next, then I watch to find out what I actually do. It seems I'm not entirely in control of the part of me that wills and acts. Watch the gap.

— Clothes do not make a man, nor does a plot make a novel. All depends on the characters, how they see and hear and think and feel; not what they do, but how they do it, who they are.

— The role of the artist is perceiving the ideal in the real — like Phidias seeing the finished sculpture in the rough rock. The world isn't broken. Rather, its imperfection is a gift, an opportunity for the artist is to continue the act of creation.

— The most difficult writing challenge is listening carefully to what the characters tell you and restraining yourself from forcing them to do and say what you want instead of what they want.

— Rereading Shakespeare is like playing a piece of music. The pleasure grows as you learn it, until you can watch it in your mind without looking at the words, like you can play music without looking at the score and can hear the music without playing it.

— Did you come to a dead end? Relax. Go for a walk. Take a nap. Meditate. Clear your mind. Then, if you're lucky, your characters will begin to talk, and you should humbly and joyfully take dictation. Later you can edit and rewrite and smooth.

— Don't get caught in the memoir trap. Truth isn't just what happened. And you can never know what really happened.

— Write for yourself. Write stories you would like to read. Write the stories that are in your gut and need to come out.

— Maybe the purpose of novels is to introduce us to people and places we could encounter in alternative lives.

— Write because you love to write. Write no matter where you are and what you are doing.

— Enjoy it for its own sake. Never expect payment. If money comes, great. But that should never be your motivation.

Immediacy in Fiction

Reality consists not just of what has happened and what will happen, but also of what might have happened. In the present, there is a cloud of possibilities (as in quantum physics) that will be resolved when a decision is made or an action is taken. This reminds me of how, in chess, after every move a new set of next moves is revealed in a branching pattern, as Kotov observes in *Think Like a Grandmaster.* In my favorite novels, the action is enriched by such a cloud of potential. A moment seen from such a perspective is much more interesting than an account limited to actual events. Readers should feel like they are present as events unfold and decisions are made.

The Unwritten Contract Between Novelists and Readers

The novelist should establish the rules of the narrative world from the start. In fantasy and scifi, that world can radically differ from the everyday world, but the rules should be clear enough for readers to extrapolate and anticipate consequences. In historical novels, the author should make the values and mores of that time clear, as well as the historical context. And authors of contemporary novels who wish their work to be enjoyed in the future should make explicit the values and mores of the present time so future readers will be able to engage.

Once the world is established, the novelist should not introduce changes in the rules that do not derive from the actions of the characters. Events should flow naturally from previous events and should have predictable consequences. Above all, the climax should not feel arbitrary. It should be the natural result of what came before. As a corollary to that, details should be significant, not random. They should affect the story and help attentive readers to accurately anticipate. The author can employ misdirection and sleight of hand, can make the reader work to discern the direction of the story and who did what and why. But at the end, the careful reader should be able to recognize how the pieces fit together, leading inevitably to the conclusion.

In return, the reader promises to read closely, paying attention to the rules and to the details, imagining what it would feel like to live in such a world and to find oneself in such a situation. And if, given that investment of effort, the reader finds the story satisfying, the reader has an obligation to spread the word.

Binge-Reading Shakespeare

We binge watch. Why not binge read?

When you change the context of writing, you change its meaning. To binge watch a TV series is to experience the episodes together, as if they were a single work, to enjoy them in a new way. In the old days, the only choice for watching series was broadcast television. Typically, 22 episodes constituted a season, and the episodes were broadcast one per week, with the time slots for the rest of the year were filled with reruns. It was a stop-start experience, often with cliff-hanger stories to lure viewers to back.

The advent of video recorders changed that experience. You could save episodes and watch them at will. You could rent or buy. You were no longer constrained by the schedule of the network or local station. You could fast-forward past commercials. You could pause. You could rewind and rewatch. You were in control.

Then came cable with video on demand and DVRs, giving you even more convenient control. Programming to record was far easier.

Now with streaming, you don't have to plan ahead. You can watch one episode after another, from the first episode of the series through the last one, without commercials. Watching in that mode, you can get deeply involved in the story. You can identify with the characters and see the actors growing up and aging — like time-lapse photography, where what normally takes days or months or years unfolds for you fast enough for you to enjoy the spectacle of change. Or you can choose to watch in stop-start mode to suit your personal schedule and lifestyle.

My favorite instance of binge watching is *Newsroom* by Aaron Sorkin, which originally aired on Showtime from 2012 to 2014, twenty-five episodes spread across three seasons. Viewed in its entirety, it has a beginning, middle, and end. While each episode is satisfying in and of itself, the series as a whole is a single work of art, deliberately written to be experienced that way.

Getting warmed up to write an historical novel set in the time of Shakespeare (S*hakespeare's Twin Sister*), I decided to binge read his complete works, one play every day or two. That's 38 plays, written over the course of about 19 years. It was a delightful experience.

I read those plays aloud to get a feel for the rhythm. As I became familiar with the vocabulary and the syntax, I didn't have to go running to the footnotes all the time. His language began to feel natural rather than alien, as I became familiar with stock phrases and images and allusions, as well as the range of reactions of characters experiencing love, jealousy, hate, vengeance, temptation, ambition.

The histories, in particular, make much more sense read together. The complexities of genealogy and royal succession fell into the background as I became familiar with them, freeing me to focus on the characters and the spectacle and the pageant.

Now I'm tempted to binge-read the complete works of other authors, sets of books that gain from being read one after the other: Balzac and Zola, who deliberately set out to tell multi-volume stories; then Faulkner, Michener, and Somerset Maugham.

This binge-reading experience prompts me to think differently about what I write and why I write. I write what I need to write rather than responding to editors or markets. Because my novel are so personal, over time, they will probably fit together and form a single coherent story. And, for me, the main purpose of writing is to discover that story and tell it.

In Just—and Hemingway

One of my favorite poems is "In Just—" by e. e. cummings. The final lines are:

it's
spring
and
the
goat-footed
balloonMan whistles
far
and
wee

High-school textbook footnotes connect *goat-footed* with Pan, a god in Greek mythology. But for so spontaneous and natural a poem, that feels like a stretch.

After watching Woody Allen's *Midnight in Paris*, I read books about Americans in Paris in the 1920s, and books by Ernest Hemingway. I was surprised to learn that e. e. cummings was in Paris when Hemingway was there, and in *A Moveable Feast* I stumbled on the following evocation of spring:

> *In the spring mornings I would work early while my wife still slept. The windows were open wide and the cobbles of the street were drying after the rain. The sun was drying the wet faces of the houses that faced the window. The shops were all shuttered. The goat-herd came up the street blowing his pipes and a woman who lived on the floor above us came out onto the sidewalk with a big*

pot. The goat-herd chose one of the heavy-bagged, black milk-goats and milked her into the pot while his dog pushed the others onto the sidewalk. The goats looked around, turning their necks like sight-seers. The goat-herd took the money from the woman and thanked her and went on up the street piping, and the dog herded the goats on ahead, their horns bobbing. I went back to writing and the woman came up the stairs with the goat milk. She wore her felt-soled cleaning shoes and I only heard her breathing as she stopped on the stairs outside our door and the shutting of her door. She was the only customer for goat milk in our building.

Both the poem and the Hemingway passage are evocations of spring. And, by chance, both cummings and Hemingway were in Paris at the same time. So, rereading the cummings poem with the Hemingway in mind gives the words of the poem new connotations, makes it more tactile and fresh.

Today's Youth and Reading

There's a plus as well as a minus to the reading habits of the young today. Tim, my youngest, is into *fanfiction.* Those are stories written by fans of well-known plots and characters from TV, movies, and videogames. There are hundreds of thousands if not millions of fanfics that you can read for free on the Internet. Many are book length. The authors have no expectation of ever being paid for this work. They write for the social-sharing high of finding an appreciative audience, no matter how small.

Since he was in junior high, Tim has read hundreds of such stories — the bad as well as the good, for long hours. He's a fast reader — polishing off the equivalent of a Harry Potter book in less than a day. He emailed the authors with comments and suggestions for improvements; and for the best fanfics, he asked permission to post those stories at his own site, and then worked with the authors to make copy-editing improvements.

Meanwhile, with the rise of ebooks, far more books are being written and published than ever before, though many of them get very few, if any, readers. It costs nothing to publish an ebook in the Amazon Kindle store. There are millions of books available there, and the vast majority of those have probably been written and published in the last ten years, by amateurs, with no expectation of ever being traditionally published.

Today's cultural climate is chaotic. Schools no longer drill the western canon into students. That makes it difficult to build on and to pass on to future generations a rich literary context for allusion and enjoyment and understanding. But thanks to the Internet, I see broad-based enthusiasm for story such as the world has never seen before.

Good things are happening and will happen. The glass is seven-eighths full. Rejoice! Write those novels and poems you've always wanted to write. Pull the unfinished ones out of dusty drawers and finish them. Don't worry about convincing traditional publishers to publish them. Publish them as ebooks or post them on a website of your own. Encourage friends and colleagues to read and spread the word. You won't get rich, but you might find a few good readers. And what more could any writer ever hope for?

Reading and the Zerg

In the *Starcraft* game series, the Zerg are one of the races struggling for dominance. While there are many Zerg, they act together, more or less, as a single entity, a single horde or hive. When they capture an opponent, they *assimilate* him or her, acquiring new strength, new powers, new perceptions.

Reading books is like that assimilation.

I have a list of the books I've read since I was a fifth grader in 1958. Over those 64 years, I've read over 3700 books. I believe that thoughts and emotions of authors whose works I've read have become part of me and enriched me.

This feels like a variation on a line in Auden's *In memory of W.B. Yeats*, "The words of a dead man are modified in the guts of the living." Auden meant it ironically. He was writing from the perspective of the poet who on death "became his admirers." I'm thinking of that same phenomenon from the perspective of me as a *reader*: the words of dead men, modified in my guts, become part of me. They nourish me. They give me strength.

When Story Matters More than Fact

In 1944, Dad was a private in the US Army, stationed in Georgia — a bugle boy waiting to be shipped to the war in Europe. The day before he was due to leave, he received orders for Officer Candidate School. It turned out that his company was sent to the Battle of the Bulge. He heard that they were all captured, without casualties, and that the train taking them to prison camp was bombed by the Allies, and that then there was only one casualty — the bugle boy, the man who replaced him, died.

Some might see that as chance. But to Dad, he owed his life that to that other bugle boy. He had an obligation to pay it back, to live a life that mattered.

Dad was given a life and, also, given a belief that he had a personal destiny. And at every decision point in his life, in the back of his head was the image of that bugle boy who had taken his place, a humbling sense of responsibility, a debt owed.

I'm reminded of the final scene in the movie *Saving Private Ryan*, at the D-Day cemetery in Normandy, long after World War II. The man who was saved is standing with his children and his grandchildren. Not a word is said. But you get the sense that the man's whole life was predicated on that sacrifice and that debt.

Fifty years after the war, when Dad had retired as a superintendent of schools and a colonel in the Army Reserves, he wrote an autobiography, and I posted it on my web site. Someone from his old company, searching for mentions of himself online, found the book and emailed Dad. The bugle boy hadn't died in the war. With the help of a group of veterans, Dad got in touch with him, and they shared life experiences by email.

Then a year later, the officer who took command of that company after Dad left for OCS and before the Bulge, chanced upon Dad's autobiography and emailed him a detailed account of what had actually happened. In fact, more than half the men in the company had died in that battle.

The original story gave Dad a sense of debt and destiny, but the facts were very different. The story he believed for so long mattered more than the facts. It gave shape and meaning to his life — a rare gift, far more important than mere facts.

The Mothers of Fact

(This is an excerpt from my fantasy ***The Lizard of Oz****, in which an elementary school class on a field trip goes to the Underworld. You need to stand under the world to understand it. There are many levels of understanding.)*

As soon as the class got ashore in the Underworld, Kathy said, "Why, I've never seen such pretty clothes in all my life. Could you please teach me how to make clothes like that?"

One of the three old ladies who were spinning and sewing said, "As a mother of fact, that could be very difficult."

Mr. Carroll introduced them, "These are the Mothers of Fact: Miss Hap, Miss Fortune, and Miss Take."

Kathy said "I'd like to learn to sew like that?"

"Sew what?" asked Miss Fortune.

"Sew pretty clothes like you're making."

"Those are very special clothes. They're costumes for our spring fête."

"Fate? What's a fate?" asked Kathy.

"Oh, that's a party. The way we do it, it's a masquerade party, and everybody wears pretty costumes and acts out silly parts. Our job is to make the costumes."

"Can I help? Please? Pretty please?" Kathy pleaded.

"Well, I'm afraid it's probably beyond you; but if you want to try, here's a needle and thread."

"But what can I use for cloth?"

"Use the fabric of time," answered Miss Fortune. "That's what we use."

"But..."

"Once you get into it, it's really quite simple, nine times easier than regular sewing — just a stitch in time."

Kathy felt silly sitting there with a needle and thread and no cloth; but she would have felt even sillier to ask again, so she just pretended she was sewing. The other kids gathered round her and stared.

"What are you doing, Kathy?" asked Mark.

"I'm sewing, silly. Can't you see?" she answered.

"But you don't have any cloth. How can you sew without any cloth?" he asked again.

"I'm just stitching time," she said.

Miss Fortune confirmed, "Yes, and she's doing a fine job of it. She'll soon have it all sewed up."

Miss Hap added, "Why that's lovely, perfectly lovely. Why that's finer than anything we've ever made. That's a very special costume. Fit for a king."

"For an emperor," said Miss Fortune. "That'll be the emperor's new clothes."

Kathy wasn't sure whether they were just being nice, or if they were making fun of her, or if they meant something she didn't understand.

Donny said, "You mean emperors don't wear anything at all, not even underwear?"

Kathy giggled and whispered to Gaynell. Gaynell giggled and whispered to Kathy.

But Miss Fortune said "There's a very special fiber for making it visible. Yes, moral fiber. The emperor has to supply that himself. It's indecent for an emperor to go around with no moral fiber."

Mark asked, "What's moral fiber?"

"Cotton grows on some plants; wool grows on some animals; and moral fiber grows on some people. They're a rare breed."

"I'd like to buy some moral fiber," said Kathy.

"Well, you don't see plants buying cotton or animals buying wool, do you? They've got to grow it themselves. Well, people can't buy moral fiber either. They've got to grow it. It grows on you. Till you're all grown up."

Mark said, "Well, Miss Osborne's a grownup. She must have some."

Everybody looked at Miss Osborne, and she blushed.

Donny said, "I don't see anything."

Miss Osborne blushed some more.

But Miss Fortune explained, "Just give her time, and it'll show. Yes, matched with the right time, moral fiber can be quite beautiful — bright red and blue and green. Really very becoming. Becoming even more beautiful."

Perhaps *The Iliad* and *The Odyssey* Commemorate a Loss Rather than a Victory.

National epics often commemorate losses, not victories. *The Song of Roland*, *The Song of Igor,* Arthurian legends, *El Cid.* In recent times, this trend is echoed in *The Star-Spangled Banner* (written after the British sacked Washington DC in the War of 1812), the tales of the fall of the Alamo, and Civil War books like *Gone with the Wind.*

Perhaps that was the case with *The Iliad* and *The Odyssey* as well.

We know that between 1100 and 800 BCE Greece had waves of migration and invasion by unknown people from the east and north. Perhaps Troy was overwhelmed by such a migration/invasion. Perhaps the invaders took the entire coast, from Troy southward, and eventually took the Aegean islands and mainland Greece as well. Ithaca, to the west of Greece, may have been the last holdout.

Bards told tales which reversed the outcome, as if Greeks had conquered Troy. Such tales raised a sense of national pride among the defeated. The stories they told were not history. Rather they were fictitious compensation for loss. Hence the importance and durability of those tales.

Rather than stir up thoughts of rebellion and revenge (like *Remember the Alamo*), these stories emphasized the role of fate and the gods, fostered pride in the legendary feats of forefathers, and ended (in *The Odyssey*) with god-sponsored reconciliation and peace.

Homer and the Myth of the Author

When we enjoy a novel, we perceive it as an entity complete in itself. We don't read early drafts or check the comments and corrections of beta readers, developmental editors, and copy editors. We don't compare the final text with what came before and try to construct a new text.

We don't think of a published novel as the work of a team of dozens (despite an author's detailed acknowledgments). Regardless of how the work of fiction came to be, we view it as if it were the creation of a single author.

That feels natural because the characters come alive for us; they feel consistent and real. We visualize what they do and imagine what they would do in different circumstances. There's an inner logic to the characters — their speech, their thoughts, their actions — that was clear to the people involved in its creation and is now clear to audiences. Violations are like striking the wrong key on a piano or playing a piano that's out of tune.

A story with living characters is a meme, with a life of its own. The identity of the author is irrelevant except as a label, making it easy to find the work in a library or with a search engine.

Such is the case with *The Iliad* and *The Odyssey.* Regardless of how and when these stories were composed, they passed through many minds and hands, both orally and in written form. And they were translated many times. This is the reverse of the old truism about too many cooks. These works have gotten richer, thanks to the cumulative work of editors, translators, interpreters, and creative readers over the course of nearly three thousand years.

If you love these works, if they have become part of the web of associations that defines who you are and how your think, then you can say, echoing the followers of Spartacus, *I am Homer*.

Odysseus and the Centaurs

At the end of *The Odyssey*, in preparation for the bow contest for the hand of Penelope, Telemachus digs a narrow trench in the dirt floor of the banquet hall. Then he sets up axe heads, lining up the holes in a row. The challenge is to shoot an arrow so that it passes through all the holes. The holes are, at most, a couple feet off the ground. It's near-impossible to lean over far enough to make such a shot. But Odysseus does so, successfully, while sitting on a stool, bending down, close to the ground.

Using a bow at such an angle is not normal, except on horseback. Visualizing that scene, I was reminded of the riding style and archery of medieval Cossacks. But in *The Iliad* and *The Odyssey*, warriors did not ride on the backs of horses. Rather, horses were used to pull chariots. There was no cavalry.

I suspect that the era portrayed in the Homeric epics was a time of transition. Immigrants from north of the Black Sea, in the area now known as Ukraine, that later became the home of the Cossacks, were moving into Greece and assimilating with the local population. These newcomers were renowned for their bareback horsemanship.

When the Conquistadors arrived in Mexico, the Aztecs, who did not have horses, thought that rider and horse were a single creature. I imagine that in ancient Greece the newcomers from the north paired with their horses seemed like single creatures as well. Hence the concept of *centaurs*.

Coexistence between the Greeks and the centaurs was problematic. The centaur Chiron served as teacher to both Heracles and Achilles.

Another centaur, Nessus, caused the death of Heracles. And at the wedding of the human Pirithous, a bloody battle broke out between centaurs and humans when the centaur Eurytion tried to abduct the bride.

The bow used by Odysseus in the contest with the suitors had once been owned by Eurytus (different but similar name), who taught Heracles archery. By another account, Eurytus set up an archery contest with his daughter as the prize (similar to the bow contest in *The Odyssey*). When Heracles won that contest, Eurytus reneged on his promise and refused to give him his daughter. In revenge, Heracles stole a dozen horses from him.

I imagine that the style of archery used by Odysseus in the contest echoes the style of the northerners. The subtext is that Greeks, not just centaurs, can fight that way, so that, in the future, survival in battle will depend on that new style.

Jocasta, Antigone, Iphigenia, Electra, Helen, and Penelope

In Greek tragedies that deal with the families of Agamemnon and Oedipus, as well as in *The Iliad* and *The Odyssey*, the action centers on a woman whose husband (were she to marry) would have a claim to the throne.

Oedipus became king of Thebes by marrying the previous king's widow, who, unbeknownst to him, was his mother. In addition, he didn't know he had a claim to the throne as the only son of the previous king, whom he had killed by accident. (*Oedipus*) When Oedipus was blinded and exiled and Jocasta committed suicide, his sons, Eteocles and Polyneices, succeeded him. At first they alternated who would reign, then both died fighting one another for the throne. (*Seven Against Thebes*) Creon, the uncle and also brother-in-law of Oedipus, filled the power vacuum. He then insisted that Antigone and Ismene, daughters of Oedipus, not marry. He enforced that edict even though his own son Haemon was engaged to marry Antigone. Creon went so far as to cause the death of his own son. While it is not so stated in the play, if Haemon had married Antigone, he would have had a claim to the throne. (*Antigone*)

In Argos, Agamemnon's right to rule was tenuous because of unresolved tensions between his father, Atreus, and his father's twin brother, Thyestes. They had contended over the same woman, Aerope, and whoever had her ruled Argos. To solidify power against his brother, Atreus killed the sons of Thyestes and served them to Thyestes as a meal. When it was revealed that Thyestes had eaten human flesh, he was exiled for such a sacrilege. But Atreus was not sanctioned for the murder of his nephews nor for having served them as a meal. Following the advice of an oracle, Thyestes then mated with his own daughter in the hope that their offspring would grow up to wreak vengeance on the House of Atreus. That son, Aegisthus, mated with Clytemnestra, the wife of Agamemnon and queen of Argos, while Agamemnon was away at the Trojan War. When Agamemnon returned victorious from Troy, Aegisthus and Clytemnestra killed him. Then Aegisthus married Clytemnestra and became king. (*Agamemnon*)

At the beginning of the Trojan War, Agamemnon's claim to the throne and his position as the overlord of the Greeks was tenuous not just because of the family entanglements mentioned above, but also because he had a daughter of marriageable age. Should that daughter, Iphigenia marry, her husband would have some claim to the throne, and the most likely suitor was Achilles. Before setting sail for Troy, Agamemnon lured his daughter to the muster point, Aulis, presumably to marry her to Achilles. Instead, he sacrificed her. (*Iphigenia at Aulis*) By another account of the same story, she was miraculously saved by Artemis and transported to a distant land where she served as a virgin priestess. In both versions, she did not marry, and hence Agamemnon's right to rule remained unchallenged. (*Iphigenia at Tauris*)

After Aegisthus and Clytemnestra killed Agamemnon, Aegisthus insisted that neither of Agamemnon's remaining daughters, Electra and Chrysothemis, could marry. Eventually, Orestes, their brother, returned from exile and, with Electra, killed their mother and Aegisthus. After complex legal arguments about revenge and matricide and the right to rule, Orestes became king and Electra was allowed to marry. (*Libation Bearers, Eumenides*)

Before the Trojan War, suitors gathered contending for the hand of Helen, daughter of King Tyndareus of Sparta. As soon as Menelaus married her, he became king, despite the fact that Tyndareus was alive and healthy. By the unspoken principle that a woman could be the symbol of the right to rule, Paris' elopement with Helen gave him possession not just of a beautiful woman, but also of the woman who represented the right to the throne of Sparta. Helen's abduction was a matter grave enough to enlist the help of all those who had been rivals of Menelaus for her hand, who, together, raised an army to take her back by force. (*Iliad*)

By an alternate version of that story, Paris took Helen to Egypt, where she stayed while he returned to Troy, and Menelaus found her there after the war. By that version, Menelaus' claim to the throne was never really in doubt. (*Helen*)

In Ithaca, during the long absence of King Odysseus, more than a hundred suitors gathered to contend for the hand of his wife, Penelope. Ten years after the end of the Trojan War, he was presumed dead, and it is a matter of doubt and debate who would succeed him — either his son Telemachus or whoever might marry Penelope. (*Odyssey*)

In all these dynastic legends, the right to rule was contingent on marriage to or possession of a particular woman. While the stories seem to have originated as examples of legal issues related to laws of succession, they were perpetuated because of their dramatic potential. Aeschylus, Sophocles, and Euripides scrupulously preserved the legendary details. And hence in their tragedies, the possibility of succession by marriage put women and love and jealousy at center stage, together with greed and lust for power.

Literary Periods and Man's Changing Image of Himself

In Medieval times, in common belief, the seasons changed, and people grew up and aged, but human nature and the nature of the universe were immutable. Man, heaven, and earth were all connected to one another and to God. And every man derived meaning from that order. *(Dante, Chaucer, Aquinas)*

In the Renaissance, as advances in science revealed mechanical laws governing the physical world, the spotlight shifted to man himself. God didn't go away. Rather, God was put in parenthesis. Man's physical body and his reasoning ability became the focus of learned attention, the source of man's dignity and meaning. *(Da Vinci, Shakespeare, Descartes)*

As greater attention led to greater understanding of the workings of the human body and mind, the spotlight then shifted, in the Romantic period, to the realm of non-rational thinking — intuition, emotion, poetic sensitivity to Nature. *(Wordsworth, Coleridge.)*

Advances in science then made Nature seem less mystical, less fraught with meaning, and the spotlight of great literature, philosophy, and science moved to the unconscious, the irrational. *(Dostoyevsky, Conrad, Freud)*

As psychology revealed that the unconscious was governed by mechanical rules, the focus shifted away from man as an individual, to entire cultures and to the collective consciousness. (Joyce, T. S. Eliot, Frazer, Jung, Faulkner)

As science and philosophy concluded that the viewer is part of the physical system, and that there are limits to what human consciousness can understand, the spotlight moved again. Literary works celebrated characters who willfully and heroically distorted their own perception. *(Beckett, Sartre in* ***Nausee****, Camus in* ***L'Etranger****, Kesey in* ***Cuckoo's Nest****, Ginsberg in* ***Sunflower Sutra****.)*

That period abruptly came to an end as it became clear that drugs could mechanically produce the same effects. And over the last sixty years, we've seen numerous serious works that hint at a realm of order and understanding lurking behind the everyday mechanical world. Sometimes that echoes the perspective of the medieval period, suggesting that there really is a cosmic order that connects everyone and everything, but that is beyond our understanding. (*Pynchon, Eco, Borges, Lem, Stoppard, Neal Stephenson.)*

Part 6:
The Double-Edged Impact of Technology

The Importance of Taking the Easier Path

A friend of mine bemoans the invention of GPS. Relying on GPS, people lose their sense of direction and their ability to use maps. He sees that as an impoverishment of our spatial awareness. He also bemoans the availability of the information resources of the Internet: people don't remember facts when they can easily look them up. He hearkens back to what was lost culturally with the invention of the printing press, and the decline in the ability to memorize with the invention of writing. He sees it as a moral failing to take the easy path made possible by advances in technology.

My view is that we are programmed to take advantage of every opportunity to do more with less, to not waste effort or memory space unnecessarily, to follow the path of least resistance, like water flowing down a hill. When an easier way to do something becomes available, we have a strong inclination to adopt it and forget the old way. That inclination has repeatedly been important for the survival and advancement of humankind.

Admittedly, our increasing dependence on technology puts us at risk should the electrical and electronic underpinnings of modern civilization vanish (through such disasters as solar flares or nuclear war). But should that happen, we'll adjust to the new reality and relearn what we need to relearn. In the meantime, we'll advance far more quickly both as individuals and as a society by supplementing our natural abilities with the use of technological innovations.

The Curse of Moore's Law

In accordance with Moore's Law, the performance of computer chips has been doubling every two or three years since the 1970s. Three years from now, you'll be able to buy a computer with the same power for half the price or a computer with twice the power for the same price. This phenomenon has been driving economic growth, increases in productivity, and social change.

The predictability is due in large part to the relationship of the physical size of a circuit to its speed. Electricity moves at the speed of light, which means (as computer pioneer Grace Hopper often pointed out) that it travels about a foot every billionth of a second. If you cut the size of a circuit in half, you double its speed and hence its performance, without the need for any other innovation. And the use of ever more powerful computers accelerates the development of the next generation of computers.

The pace of this advancement is slowing now as technology reaches physical limits, though major innovations, like quantum computing, might keep the trend going for a while.

So what is the *curse*? We have experienced nearly fifty years of incredible progress in technology. As individuals, we have become used to the consumer electronics products we buy becoming obsolete quickly. Rather than fix a broken device, we buy a new more powerful replacement. We take such progress for granted. And the economy depends on this trend for increasing productivity and lowering the cost of technology products. When this progress comes to an abrupt halt — as it eventually must — the repercussions are likely to be disastrous.

Thoughts on the Significance of Ebooks

A book-as-artifact, like an antique, has value based on its rarity. A book-as-content has value only for the words and what they mean. Thanks to ebooks, book-as-artifact has been severed from book-as-content.

Books-as-artifact will be valued forever and will increase astronomically in value as they become more rare. Ebooks have no rarity. If one copy exists, millions of copies can be created instantaneously and distributed around the globe at little or no cost.

Many of the books now available for free or at little cost as ebooks used to be rare. Now they will be readily available forever, barring the possibility of another Dark Age that eliminates the technology that we now take for granted. Book burnings, à la Hitler and Savonarola, are no longer possible in this electronic environment.

At first *The Iliad* and *The Odyssey* only existed in mind and memory, and few knew them in their entirety and shared them with others by recitation. Probably hundreds of years later, writing made them more widely available. It was tedious and expensive to make copies, but it could be done. The artifact nature of written books made it possible for books to passed on from generation to generation, independent of faulty human memory. Of course, those artifact books were subject to wear and destruction. But there were always people who valued them enough to make fresh copies or to pay others to do so. Of course, there were mistakes in judgment and there were natural disasters and human disasters, but many books survived for hundreds or thousands of years.

Electronic books free the content from the artifact — like it was before books were written down. The book resides in memory once again, only the memory is electronic instead of human. Today, the content of all the literature of the entire human race can fit on a backup disk drive you can connect to your PC, and soon, with the predictable doubling of computer power, all of literature will fit on a flash drive, and then on a gadget as small as an earring, and then will be available to all devices everywhere instantaneously from cloud storage.

When Michael Hart founded the Gutenberg Project to make public domain books available for free to everyone, he compared the invention of ebooks to the invention of the printing press. But the change was more revolutionary than that. It was comparable to the invention of writing.

The Internet and The Human Spirit

We are bombarded daily with news of hacking, identity theft, and scams. We see our nation becoming ever more polarized as people ignore non-partisan balanced reporting of news and instead focus on the web sites and cable channels that agree with their pre-set biases. The communities of common interest that the Internet made possible have become havens for bias and bigotry and closed mindedness.

Innovations developed for one purpose often end up producing opposite effects. Fire can warm and cook, but it can also destroy.

Misuse of the Internet should not blind us to the Internet's potential.

We are shaped by our environment, and the Internet and social media constitute a new environment shaping everyone from toddlers to retirees, enabling new behavior, new ways to learn, and new ways for people to interrelate.

The human potential for mass destruction existed before the invention of the weapons that made it possible. The potential for people to submerge their identity and their individual reason in large-scale crowd hysteria existed before the invention of mass communication media.

But now, thanks to the Internet, we are learning that man also has the potential for large-scale reasoned discourse. Thousands or even millions of people can arrive at mutual understanding through dialogue. Ideas can be spread instantaneously in global forums where they can compete on the basis of their merit, and people around the world can work together on large-scale projects, just as they can play massively interactive videogames. In other words, the Internet reveals positive aspects of human nature that were never seen before.

Before the Internet, large numbers of people could work together only when regimented, disciplined, and controlled by a central authority. Without such control, large groups of people were dangerous and volatile, sometimes turning into irrational mobs. And mass communication — communication from one to many — could induce crowd hysteria at a distance, and one person's nightmare could be projected onto many.

By enabling many to communicate with many, directly, without central control, the Internet reveals unexpected human potential for collaboration and community. I hope that today's divisiveness is a transient anomaly.

Time for Reflection

Have you ever been to a football game where the audience holds up cards that together spell a phrase or compose a picture? Imagine a much larger group of people forming a picture that can be seen from satellites. Imagine that the people are holding large panels of cardboard and one side is covered with highly reflective material, like aluminum foil. And they hold it, bright side up, horizontal at noon on a sunny day. Imagine a million or more people doing that at noon their local time.

The act itself, the demonstration, is an act of protest and solidarity against the deterioration of the Earth's climate. It's a statement of unity, visible from satellites, perhaps even from the moon.

Remember the physical mechanism behind the movie *The Day After Tomorrow,* in which sunlight, rather than being absorbed by ground and plant life, is reflected by snow and ice, rapidly cooling the Earth and bringing on an ice age.

Measure the temperature of the atmosphere before and after such demonstrations to determine if there is a measurable change in temperature. A temperature change could serve as proof of concept for a project to combat global warming by placing reflective panels on rooftops and across desert areas, designed to change angle and direction in response to the position of the sun.

Imagine using slats instead of panels (like venetian blinds) that could generate electricity. Such an approach might mitigate or even reverse global warming in a controllable way, at the same time generating electric power which would reducing dependence on fossil fuels.

Part 7:
History

Yes-terday

There has always a *good old days*. The past is always simpler than the present because we know so little about it. We remember the events that shaped the world as we know it today, but we forget the branching paths of possibility that gave context and meaning to the events as they unfolded.

In the present, we are inundated by an infinite number of possibilities. But when we consider the past, the choices and the challenges seem much easier to deal with, not because they were, but because of our ignorance of what was at stake then.

The Myth of Future Shock

Many people presume that today the pace of change is ever-increasing, making it ever more stressful for us to adapt. But, in fact, the changes we experience today are fine-tuning — adding functions to a smart phone, watching TV shows and movies streaming over the Internet instead of broadcast or on DVDs. In contrast, the changes that occurred in the 19th and early 20th centuries were fundamental and disruptive. The Pony Express lasted just 18 months and then was made obsolete by the transcontinental railway. Locomotives, automobiles, electric lights, telegraph, telephone, radio, photography, sound recording, movies: an explosion of inventions radically changed everyday life.

Trigonometry and American Independence

George Washington was good at trigonometry, so good that he qualified for a job as a surveyor of frontier land. Then, because he was familiar with frontier land, General Braddock gave him the rank of colonel in the British Army and relied on him to guide the expedition to Fort Pitt. That expedition was ambushed and defeated by French and Indians. But when the Revolution began, the Continental Congress made Washington Commander-in-Chief of the Continental Army because of his military experience and his rank of colonel. And with his unique style of command, Washington won the war, despite losing every battle but the last.

If Washington hadn't done so well in trigonometry, he wouldn't have become a surveyor and wouldn't have been made commander, and we might have lost the Revolution.

The Genius of Washington

(Adapted from a chapter in my novel ***Parallel Lives****, based on passages from the history of the Revolution written by Mercy Otis Warren, the first American woman historian).*

Washington arrived in Boston in the summer of 1775, a year before the Declaration of Independence. That was after Concord and Lexington, after Bunker Hill. He had nothing to do with any of the fighting in Massachusetts.

Most members of the Continental Congress didn't want war. They were hoping and expecting that England would agree to reasonable terms. The make-shift army under Washington was a preventive measure, in case the British should decide to retaliate and punish the colonists of Massachusetts for opening hostilities. Both sides were holding fire, waiting for diplomacy to resolve their differences. Washington had a small force of untrained farmers and tradesmen, facing an army of professional soldiers. There was no telling how long the negotiations between Congress in Philadelphia and the King and his Ministers in London would drag on. It took six weeks for a message to go from one place to the other.

When Washington arrived in Boston, he learned that his forces had just three rounds of gun powder per soldier. That deficiency was a closely held secret. If the British had known, they could have wiped out the colonists with ease. Washington acted as if nothing were amiss. He had his troops build fortifications on the hills around Boston, as if he had all the ammunition he might ever need.

The local farmers had little gunpowder to spare. The southern colonies eventually sent some along. The locals did their best to tool up to produce gunpowder, but that took time. What made a real difference was pirates -- privateers -- empowered to prey on British shipping and to capture whatever military supplies they could. And in the meantime, Washington successfully carried on his charade, tricking the British into thinking that he was well supplied and they were at risk.

That was the first indicator of Washington's genius. He would have been great at poker, bluffing to get the most out of his cards, even when he had a losing hand.

Six years later, after France had entered the war as an ally of the colonists, Clinton was in command of the British forces in America, with his headquarters in New York City. Washington had a small force, consisting mostly of untrained recruits. Volunteers would come and go, heading home when they wanted to plant and harvest and deal with family matters. Now he had a small French army as well, under the Count of Rochambeau. They joined forces in White Plains, near New York City. A French squadron had set sail from Rhode Island, presumably heading for New York as well.

Clinton intercepted dispatches from Washington confirming plans for attack, and either those dispatches were decoys, or Washington, knowing they had been intercepted, changed his plans. His army was far smaller and weaker than Clinton presumed. Moreover, news had arrived that more Hessian mercenaries would soon join the British forces. So instead of attacking New York City, Washington marched all the way to Virginia in three weeks. And the French squadron in Rhode Island sailed south and joined a French fleet that was already on the Chesapeake, bringing another 3000 troops for Washington's army.

While Washington was on his way to Virginia, a skeleton force left behind kept up the appearance that the rebels were at full strength and ready to attack New York. They marched back and forth on the ramparts of forts, carrying broomsticks instead of guns. Clinton was so convinced of the imminent danger that he sent orders to Cornwallis in Virginia, to send some of his troops north to help defend New York, when, in fact, Cornwallis was about to be attacked.

Also, the Battle of Yorktown itself wasn't a clash of armies on an open field. Rather, it was fought and won mainly by shovels. The British had just 400 shovels, far less than the colonists. The colonists dug trenches parallel to the British earthwork defenses, moved up their cannon and bombarded. Then they dug channels leading closer and dug trenches parallel again so they could move their guns up again. And so on. Meanwhile the British, who were stuck on a narrow peninsula, blocked on the land by the combined American and French armies, and on the sea by the French fleet, were running out of food and ammunition. Cornwallis tried to save the bulk of his army by sending them up the river on boats, but a storm broke that up. Despairing that his army would be totally destroyed before reinforcements arrived from New York, Cornwallis surrendered.

I'm used to thinking of history as inevitable -- that what happened had to happen: it was fated. But what if nothing was fated? What if one unlikely event after another hadn't happened, like Washington having the genius to fool Clinton and march, unnoticed, all the way to Virginia? What if it was just by chance that things turned out the way they did? What if, with one small change, it could have turned out very differently? If the people involved in those events had suspected how much depended on even their smallest actions, that would have been a heavy responsibility. Crippling, probably. They might have been so afraid to do the wrong thing that they wouldn't do anything at all.

The Genius of Mercy Warren and the Perfidy of John Adams

I didn't just read Mercy Warren's *History of the Rise, Progress, and Termination of the American Revolution*, I typed it — all 1317 pages of it.

I love accounts of historical events written by people who lived at that time, as opposed to works written long after, where the importance of an event depends on its relationship to today's world view.

I had read about Mercy Warren years before when I wrote a play about her and General Burgoyne for the Bicentennial in 1976. But I had never read her history of the American Revolution. The length was daunting, and the only available edition of it — a photographic facsimile of the 1805 edition — was very difficult to read. The old-style typography (the letter *s* looked like *f*), combined with the outdated spelling and punctuation (sentences that ran on for many lines), were hard on the eyes. I could force myself to decipher a paragraph or two, but then my mind would wander. Typing it forced me to pay attention to every word.

Why should I want to pay such close attention? Here was a little-known first-hand account of the American Revolution. It could impact our image of our nation's origins and destiny. I wanted to make it widely available in readable form. And thanks to the Internet, it wouldn't take years or money to do that. All I needed to do was type it, modernizing the typography and punctuation and editing for readability. I did so and made it freely available to everyone through my Web site (seltzerbooks.com/warren).

The author was a woman, writing at a time when it was very rare for women to write history. Yes, her style was pompous, mimicking the rhythms of Burke's speeches and Gibbon's history of Rome, and making generalizations about the nature of man and war and politics, rather than providing all the raw data and first-hand observations which she had at her disposal. But her personal voice came through, becoming louder and clearer toward the end of the war, and virtually screaming in the final volume. There she expressed her concern for the future of the young republic. She dared to criticize Washington's dependence on his military cronies and harshly accused her life-long friend, John Adams, of having lost faith in democracy and favoring monarchy. In 1801, just after the end of Adams' term as president, she wrote those words with passion, fearing civil war. The divide was becoming so great between those who believed in the principles of equality stated in the Declaration of Independence and those who had a nostalgic reverence for the aristocratic style, titles, and pomp of European courts. She foresaw, with dread, the special status of the super-rich two hundred years later. "...it may be feared that the spirit of avarice will be rendered justifiable in the opinion of some as the single road to superiority."

In her day, the primary political divide was between those who, like herself, believed in the Republic as a moral, even a religious necessity, and those who saw it as a temporary expedient. To Mercy, the American experiment in democracy of the people, for the people, and by the people was a beacon to the world, a shining example that could lead all peoples to free themselves from the tyranny that kept many in misery, poverty, and slavery.

Men like Adams responded to the French Revolution with fear and loathing. From that blood-fest they concluded that democracy was flawed, that it was at best a temporary solution. Mercy cites with disdain and disappointment Adams' book *Defense of Their Constitutions,* which "drew a doleful picture of the confusion and dissolution of all republics." She did not mention the undeclared war which Adams had waged on the seas against republican France. She did not mention his Alien and Sedition Acts, which revoked much of the Bill of Rights, purportedly for public safety in a political atmosphere resembling that of the McCarthy era 150 years later. She focused instead on what she considered even more insidious than that, as if he had taken a mistress — monarchy/aristocracy — while still married to his wife — republicanism/democracy.

Mercy admitted that Adams was not alone in this betrayal: "The confusions introduced into that unhappy nation [France] by their resistance to despotism and the consequent horrors that spread dismay over every portion of their territory have startled some in the United States... and shaken the firmness of others, who have fallen off from their primary object and by degrees returned back to their former adherence to monarchy..."

Though reluctant to attack her old friend, she felt it was her moral duty to do so — not just to set the record straight, but to alert the young Republic of the danger and to help nudge it in the right direction, so it would have a chance to survive, to grow, and to thrive in a world dominated by monarchs and dictators.

"The veracity of an historian requires that all those who have been distinguished, either by their abilities or their elevated rank, should be exhibited through every period of public life with impartiality and truth. But the heart of the annalist may sometimes be hurt by political deviations which the pen of the historian is obliged to record. Mr.

Adams was undoubtedly a statesman of penetration and ability; but his prejudices and his passions were sometimes too strong for his sagacity and judgment."

She blamed his five-year sojourn in England, as a diplomat, after the Revolution, for having led to this anti-democratic change in his convictions: "...unfortunately for himself and his country, he became so enamored of the British Constitution and the government, manners, and laws of the nation, that a partiality for monarchy appeared, which was inconsistent with his former professions of republicanism... After Mr. Adams's return from England, he was implicated by a large portion of his countrymen as having relinquished the republican system and forgotten the principles of the American Revolution, which he had advocated for near 20 years."

Ironically, despite her faith in republicanism, when Mercy waxed religious and invoked the name of God, she used paternalistic and monarchic images like *the kingdom of Christ.* Perhaps that was due to her familiarity with King James' diction. But perhaps she had not yet worked out how God's kingdom could be republican.

Mercy made it clear that the Revolution impoverished and greatly disrupted the lives of nearly everyone, not just the soldiers. The drama didn't end when the last shot was fired. She emphasized the economic side of the devastation, the hyperinflation with soldiers and suppliers paid in paper money that became worthless; and the ruthless activities of speculators, buying up promissory notes from patriots for practically nothing and then getting payment from Congress. It wasn't a pretty picture. The brave, cold, hungry, sick soldiers mutinied more than once. Some were executed. Everyone didn't live happily ever after. But the Republic survived, stumbled forward, and tried to find the true path to a destiny that could change the world forever.

Taking a fresh look at Will and Ariel Durant's *Story of Civilization*

The Durant series includes:

- Our Oriental Heritage, 1935
- The Life of Greece, 1939
- Caesar and Christ, 1944
- The Age of Faith, 1950
- The Renaissance, 1953
- The Reformation, 1957
- The Age of Reason Begins. 1961
- The Age of Louis XIV, 1963
- The Age of Voltaire. 1965
- Rousseau and Revolution. 1967
- The Age of Napoleon, 1975

As the title says, Will and Ariel Durant tell a story — one whopping story, from the beginnings of civilization up to the 19th century. This isn't academic history. It's entertainment for the millions. There's no need to read it from beginning to end. If you try to, you'll never finish. But you can read a chapter here and a chapter there, following the threads that weave in and out from volume to volume or following your current interests.

The Durants tell this story from a point of view. That's natural with history, though academic-style histories often mask their bias. They don't pretend to be objective. They call it as they see it, with strong and well-expressed opinions. For me, much of the charm and delight of the work comes from those opinions and that excellent writing style. Gibbon, too, was very subjective, with a delightful style, including judgments that sum up individuals, countries, and periods. But Gibbon only dealt with a small subset of the vast topic that Durants took on.

Sometimes the Durants characterize a long complex career with a few incisive sentences. For instance, "Charles V was the most impressive failure of his age, and even his virtues were sometimes unfortunate for mankind." *Reformation*, p. 642. Also, about Christian II of Denmark, "Christian fled to Flanders with his queen, the Protestant sister of Charles V; he made his peace with the Church, hoping to get a kingdom for a Mass; he was captured in a futile attempt to regain his throne, and for twenty-seven years he lived in the dungeons of Sonderborg with no companion but a half-wit Norwegian dwarf. The paths of glory led him with leisurely ignominy to the grave (1559)." *Reformation*, p. 628. There's much here to ignite the imagination of a novelist.

Sometimes the Durants succeed in capturing in just a few words the crux of a situation, for instance about Loyola, in his days as a soldier in Pamplona, "Four years he spent there, dreaming of glory and waking to routine." *Reformation,* p. 906.

Elsewhere, they render cursory and eloquent judgment, for instance, about John Calvin *Reformation* p. 490, "...we shall always find it hard to love the man who darkened the human soul with the most absurd and blasphemous conception of God in all the long and honored history of nonsense."

The books in this series were published over the course of 40 years, 1935-1975, and some of the 11 volumes are over 1000 pages long. But this massive work has found its way into the hands of many people over the years, mainly as a perennial new-member enticement for the Book-of-the-Month Club. That's how I got the first volumes, back in 1959. For many middle-class, baby-boomer Americans, these books were and remain the standard historical reference work.

But reading the Durants today, I can't help but recognize how much has changed, with the dissolution of the Soviet empire and the collapse of Communism. It's only natural to tell a *story* from the perspective of *today*. Now *today* has changed, so the *story* feels dated. *The Story of Civilization* is still great entertainment and a handy reference work for checking dates and names, but the overall thrust of the narrative no longer resounds with authority.

For the Durants, the events of previous centuries were important as causes or harbingers of what in their day looked like the ultimate conflict facing mankind. They highlighted every minor event and character with any possible connection to the eventual conflict between Communism and Capitalism. While the narrative ended with Napoleon, the implication was that the story led inevitably to the Cold War issues and conflicts that prevailed when the Durants wrote.

But today, the Cold War is a distant era that we can only understand with research and effort. It's difficult to reconstruct the perspective and set of assumptions that permeated much of Western thinking for a generation, but is now gone. Today, there is no ultimate conflict. Hence, we no longer see history in Hegelian terms, with events unfolding in a single direction. We can now appreciate history as story, as the story of humankind, and it can come alive again — in many different tellings of many different episodes. I wonder what interesting and obscure events and people will now be resurrected from the junkheap of history.

Today, we can look back on the 20th century as a play in three acts — WWI, WWII, and Cold War — rather than as the culmination of all history. Only when the Ice Age ended could anyone conceive that ice was not the ultimate state of nature, that there would be other trends and cycles — some short and some enormously long. During the Cold War, reading history was like reading a story when you already knew the outcome. Yes, you could appreciate the details and the performance, but it all led to what you already knew. That was history seen through the colored lenses of the major issues of that time.

What a relief it is to live now in a time when the major issues are unknown. Today, we have the opportunity to look at the past with fresh eyes, with new undefined and shifting filters. The past is alive — not yet killed by a new orthodoxy.

History changes over time. We always view the past teleologically, as if what came before is of significance mainly in so far as it led to the world becoming the way we see it today. As the present changes, the teleology changes. Hence the need for each generation to rewrite history.

The Slavery That Was Rome, in Plautus, Terence, and Petronius

The most popular plays of Plautus and Terence focus on the role of slaves. In Plautus' *The Prisoners of War,* the prisoners, slaves by capture, face a moral dilemma: should they try to escape, breaking the code of honor, or wait to be ransomed by their families? In Plautus' *The Rope*, the beautiful young girl whom the hero loves is a slave, and he seeks to buy her from her owner. It turns out that she had been born free but was kidnapped as a child and sold into slavery. In *The Brothers* and *Phormio* by Terence, the men are concerned not about winning the hearts of slave girls but about bargaining with the procurers who own them.

While Plautus and Terence both borrowed their plots from Greek sources, they modified them in accordance with Roman slavery laws. Legal niceties are often key to the resolution. Extrapolating from those plays, what were those laws?

Slaves can have, earn, and save money. If they save enough, they can buy their own freedom. The procedure for an owner to free a slave is simple and informal. You tap the slave with your hand, turn him around, and say "Be free, henceforth." But in addition to paying their masters, slaves also have to pay a substantial tax to the Roman government to legalize the transaction.

Slaves can be trusted advisors, teachers, and companions of their owners, but they cannot plead a legal case, and their testimony is inadmissible in a court of law.

There is no obvious physical difference between slaves and freedmen. It's not a matter of race or even nationality. And record-keeping is sloppy, making it easy to kidnap children and sell them into slavery..

Prisoners of war become slaves. Debtors can sell themselves into slavery to pay off what they owe. Criminals can be enslaved as punishment. And the children of slaves are slaves.

Owners can do whatever they please with their slaves, including hiring them out as prostitutes. And there's no sense that there's anything morally wrong with the owners who act as procurers or the slaves who do their bidding.

Prosperous young men who enjoy the services of slave girls sometimes fall in love with them and then seek to buy them from their owners. Otherwise, these young men would marry as arranged by their parents — a financial transaction, with the bride's family paying a dowry. Buying a prostitute slave is portrayed as more romantic than an arranged marriage. In the comic resolution of such tales of sweet innocent love, it may turn out that the slave girl is actually from a good family, having been kidnapped as a child, and that she's exactly the one that the parents would have wanted him to marry in the first place.

An owner can execute a slave, without justification. But, surprisingly, slaves are typically portrayed as loyal to their masters. They accept their fate and their role. They are bound by a code of honor; and don't seem inclined to try to escape. They focus instead on persuading their owners to free them or on trying to earn the money needed to buy their own freedom.

By the time of Petronius, the Republic was dead and many of its institutions had changed, but not slavery. In a world in constant flux, slavery was a transitional condition. Ambitious slaves needed to be prudent — to please their owners, to save the allowances their masters paid them, and to earn additional money, in the hope of buying freedom and paying the manumission tax. Once freed, they could rise socially as their wealth increased.

In the *Satyricon*, Trimalchio, the nouveau-riche party-giver, is a former slave, as are many of his wealthy guests. One such guest came from the provinces and voluntarily sold himself into slavery, not because of debt but because he knew that the prospects for advancement as a slave in Rome were far better than those of an ordinary taxpayer in the provinces.

The conspicuous consumption of these wealthy former slaves is part of their world view. The fortunes of everyday life are as arbitrary and unpredictable as the despots who rule the Empire. With no belief in gods or moral rules or in an afterlife with judgment and retribution, these former slaves play the easy-come, easy-go game of life as best they can. They eat, drink, and make merry. They enjoy watching gladiatorial contests as confirmation of their good luck: at least for today, they aren't the ones suffering and dying. Only chance separates the onlookers from the victims.

Abolition of Slavery in Virginia in 1775

In her history of the American Revolution, Mercy Warren told the little-known tale of the British emancipation of slaves in the South. In 1775, Lord Dunmore, governor of Virginia, freed the slaves in his colony and armed them to intimidate the colonial rebels. "Irritated by opposition, too rash for consideration, too haughty for condescension, and fond of distinguishing himself in support of the parliamentary system, Lord Dunmore dismantled the fort in Williamsburg, plundered the magazines, threatened to lay the city in ashes and depopulate the country: As far as he was able, he executed his nefarious purposes. When his lordship found the resolution of the House of Burgesses, the committees and conventions was nowhere to be shaken, he immediately proclaimed the emancipation of the blacks and put arms into their hands. He excited disturbances in the black settlements and encouraged the natives bordering on the southern colonies to rush from the wilderness and make inroads on the frontiers."

Perhaps, had the British won the war, slavery would have been abolished in America four score years earlier.

The Stages of Human History

Human history is typically described in three stages: hunting/gathering, farming, and industrial. Consider an alternative grouping: hunting/gathering, farming/industrial, and techno-global.

Each of these stages has a different value system learned through the struggle for survival. The industrial stage is an extension of the farming stage. And we now find ourselves at the beginning of a new stage.

Stage one — hunting/gathering, pre-history

- Humankind is at war with beasts and the elements.
- Knowledge, skill, and experience are necessary for survival in a world over which people have no control.
- Physical strength and personal survival skills matter.

Stage two — farming/industrial, from pre-history to the end of the 20th century

- People domesticate and control beasts of burden.
- Then, with the advance of technology, they substitute machines for animals.
- Success in farming depends on weeding, killing off runts, eliminating the weak and handicapped.
- Owners of land, beasts, and other humans use values learned from farming to support industrialization and nationalization.
- Managerial skill (ability to control others) is valued on the farm, in the factory, and in government.

Stage three — techno-global, now

- Based on global communication and economy, global interdependence grows.
- Knowledge, skill, and experience are valued.
- Technology makes possible new ways of working together and living together.
- Technology enables group action and coordination without central control
- The weak and handicapped deserve equal rights.
- Individual responsibilities extend beyond the family and local district.
- Technology extends human knowledge and capabilities.
- We are all citizens of the entire planet.

From this perspective, farming and industrialization were different aspects of the same control-based value system, which lasted more than 10,000 years. And we are now at the beginning of a new age, characterized by cooperation, compassion, and interdependence.

World War II Legacy

Traumatic events can change how an entire generation views the world, leaving a strong imprint long after. For instance, we came out of World War II with three assumptions: 1) good and evil are real and identifiable, on a large scale, and some wars are just, 2) people in crowds tend to act irrationally, more like animals than human beings, and 3) the kind of rapid technological progress spurred on by the war would continue indefinitely into the future.

We came out of the Viet Nam War with a different set of assumptions: 1) good and evil are interconnected, and there are no just wars, 2) resources are becoming scarce, 3) every technological advance comes at a high price.

We see the world from the perspective of the crisis that marked our youth. And we remain on alert for a recurrence of a similar event. We are always preparing to fight the battles of the last war and hence set ourselves up to be surprised by the next generation-defining event, which inevitably will be different from the last one.

How will the current generation be shaped by today's crises: global warming, threats to democracy, and Putin's war on Ukraine?

Predictable and Disastrous Belief in Magic

When we see history repeat itself disastrously within the same culture, we presume that people failed to learn from the lessons of the past. And when we see similar disastrous events unfold in different cultures, we look for contacts between the cultures, as if the penchant for mass self-destructive behavior were a contagious disease; and without contact, the copycat events would not have occurred.

But sometimes similar outlandish and awful events occur independently, in cultures isolated from one another. Analysis of such incidents might reveal principles of human nature — not commonality of dreams and neuroses due to the collective unconscious, but rather predictable rules of human behavior that could guide international relations and public policy.

One such constellation of horrendous events occurred during the late 19th century: the Zulus in the Anglo-Zulu War of 1879 in South Africa, the rebel forces of the Mahdi of 1881-82 in the Sudan, the Ghost Dancers of 1888-92 in the American Northwest, and the Boxers of 1897-1901 in China. In all four cases, native peoples rose up against far more powerful, technologically advanced foreigners. The native culture was about to be obliterated economically as well as militarily. Their traditional way of life was no longer viable. Everything they had believed in was being undermined daily by these foreigners, who were already in control. The situation was hopeless. Then, spontaneously, in all four unconnected instances, mobs of natives, with or without leaders, attacked the foreigners, believing that supernatural salvation was imminent and that by ritual and magic they could make themselves invulnerable to the foreigners' bullets.

Timeless Scams

The letter began: "Dear Sir, Although I know you only from good references of your honesty, my sad situation compels me to reveal to you an important affair in which you can procure a modest fortune, saving at the same time that of my darling daughter."

Dated April 3, 1914, the letter was ostensibly sent by Serge Solovieff, an embezzler and murderer in prison in Spain, to my great-uncle Charles Seltzer, in his 20s, living in Philadelphia, and just starting his career as an architect. I found the letter and a related newspaper clipping in a box of his belongings when he died in 1970. I was intrigued.

The clipping said that Solovieff, a banker in St. Petersburg, had embezzled over five million rubles, murdered a compatriot in Spain, been apprehended in London, and extradited to Spain. The money was still missing. There was no date on the clipping, but the item on the reverse side was a review of an issue of the *London Quarterly* on the centenary of Tennyson's birth (1909). It's hard to imagine reviewing a magazine long after it was published, but the letter was dated 1914 — five years later.

Why would anyone keep a clipping from an English newspaper for five years in a prison in Spain, and then send it to a total stranger in the U.S., with a letter asking for help? Was this a hoax that someone tried to play on my great-uncle? What was there to gain?

In an effort to identify this Solovieff, I searched in microfilms of English newspapers from that time. I soon reached a dead end.

Thirty years later, when the Internet was available, I included the text of the letter and of the clipping in an article that I posted at my Web site. The page got included in search engines, so people looking for the name Serge Solovieff or phrases in the letter or clipping found my page and contacted me. This is a random research technique that I call *flypaper* — posting content on the Web so people with similar interests will find you and bring you the information you need when there is no way to actively find that information.

As a result of that posting, I was contacted by half a dozen people who possessed nearly identical letters and clippings that were addressed to their relatives just before or during World War I. One of them had a second letter as well. Her great-uncle took the bait. He sent a cable to Spain in 1913 and received an 11-page letter, a masterpiece of persuasive deception, detailing how to get to Spain and what to do to retrieve the fortune.

I told this story to an old friend, Ashley Grayson. He told me that variations on this scam, popularly known as *The Spanish Prisoner*, have been around for a long time. He pointed me to the movie *The Spanish Prisoner*, written by David Mamet and starring Campbell Scott, Ben Gazzara, and Steve Martin. That movie, which portrays an elaborate confidence game, includes the following passage, which is the source of its title: "It's an interesting setup, Mr. Ross. It's the oldest confidence game on the books. The Spanish Prisoner... Fellow says, he and his sister, wealthy refugees, left a fortune in the Home Country. He got out. The girl and the money are stuck in Spain. Here is her beautiful portrait. And he needs money to get her and the fortune out. Man who supplies the money gets the fortune and the girl. Oldest con in the world."

Ashley also connected this scam to the Internet-based *Nigerian Letter*. (For details, see http://home.rica.net/alphae/419coal/) I myself receive an average of two variants of this letter every day. My email inbox seems to be a magnet for such messages.

Here's how the scam operates: the target receives an unsolicited message, purportedly from Nigeria, containing a money laundering proposal. Sometimes there's a "bequest" left you in a will or someone has a lot of currency that needs to be "chemically cleaned" before it can be used. The variations are creative and numerous.

I've received copies of documents from six instances of this scam, with almost identical introductory articles and clippings but with different handwriting on all of them and with two different names for the sender. These letters range in date from 1911 to 1914. They were sent to individuals from Maryland to Washington State, who apparently had nothing in common and apparently had no previous connections with Spain. I have no clues as to how the victims were chosen or how the senders got their addresses.

My best guess is that hundreds, if not thousands of these letters were sent to random recipients, like modern email spam, and that enough of these recipients took the bait to make this a profitable venture — one to be tried again and again and to be passed along from one schemer to another. Impediments to trans-Atlantic travel brought on by World War I may have brought *The Spanish Prisoner* to an end. In any case, it's probable that the perpetrators were never caught.

I wouldn't be surprised if there were similar scams in Roman times. The vendors outside the Colosseum in Rome today probably resemble the vendors who preyed on tourists at the same spot 2000 years ago.

And there were probably sketch artists at the Pyramids in ancient times, drawing pictures of tourists on camelback just as their descendants take photos today.

To me, the commonality and continuity of human vulnerability to scammers and peddlers is reassuring — it reinforces my feelings of community with humanity through the ages.

Part 8:
Business and Product Ideas

Million-Dollar Opportunities

When you are tired and frustrated and you know that the hassles you are going through are unnecessary, instead of venting, try to come up with a solution. Every hassle you face is a million-dollar business opportunity.

Have you ever been sitting on a toilet in an airplane bathroom when the captain announced that turbulence was coming and everybody should fasten their seatbelts? Have you ever seen a seatbelt in an airplane bathroom? Why not?

At a baggage carousel in an airport, have you realized that your suitcase has been going round and round, and you could have left long ago if you had quickly identified it? I used to have a similar problem in large parking garages, forgetting where I had left my car and walking up and down hoping to spot it. Now that problem is solved with a button on my electronic key that triggers a beeping noise from my car. Why not do the same for a suitcase, generating a distinctive sound, perhaps even saying my name?

Have you ever been embarrassed overhearing personal matters or unseemly language at a nearby table in a restaurant? Or have you wanted to avoid having your conversation overheard by restaurant neighbors? At home, you might use a white-noise generator to mask sounds from your neighbor's apartment. Why not have white-noise generators and partitions available in restaurants perhaps for a nominal fee?

Electronic ankle bracelets alert the police when the wearer goes outside the permitted area. For greater effectiveness in virtual imprisonment, why not add the ability to jam electronic signals, to prevent the wearer from using a cell phone or connecting to the Internet.

Devices are now available that can translate spoken speech and generate voice output, making it possible for people who speak different languages to converse. Why not develop a smartphone app that can recognize and interpret culture-specific body language?

Sometimes loading and unloading airline passengers takes 15-30 minutes, with everyone going in and out of one door. Large aircraft typically have a door at the back as well as the front, but only the front door gets connected to the movable passageway that leads to and from the terminal building. An airport that allowed two ways in and out instead of one could cut boarding/unboarding times, pleasing time-conscious passengers and allowing the airport to handle more flights with the same facilities.

Graffiti is everywhere. Removing it or painting over it is expensive and futile: new graffiti soon takes its place. Corporations could hire graffiti artists to embed corporate logos in their designs, and the artists would then police their own work.

Typically, at a Chinese or Japanese restaurant, you're given chop sticks to eat your rice and your main course, and a plastic spoon for your soup. Why not make hollow chop sticks that can double as straws?

Imagine a parking meter that is also a charging station for electric cars. While you are parked, your car recharges. You pay for parking and also pay for the power. Wide-spread availability of such charging stations would increase the appeal of electric cars.

In double-blind studies, where both groups believe they are receiving an experimental medication, but one group is not, the group receiving the harmless substance typically shows measurable improvement, sometimes almost as much improvement as the people getting the real thing. Apparently, if we think we'll get better, we do, or at least the symptoms subside. So why not sell a *placebo pill*? Label it and advertise it, making no clams that it is real medicine. Clearly indicate that this will not cure anything. But explain that it may provide fast relief for symptoms if you believe that it will give you relief.

Have you ever been late for work because the noise of your air conditioner masked the sound of your alarm clock? Why not build alarm clock features into air conditioners? When the designated time arrives, the air conditioner shuts off and the alarm sounds.

Missed Business Opportunities

Visiting Cambridge University in the UK, I learned that the individual colleges operate with separate budgets and typically get about a third of their funds from tuition, a third from endowment, and a third from operating a bed-and-breakfast and catering service when school is not in session. Colleges in the US should do the same.

Then, in London, trying to get tickets to shows at the Globe and the National Theater, I was frustrated to find out that the one was closed getting ready for the next show, and the other was sold out. Why don't they sell tickets to their rehearsals? That way, they could make more money and also make more theater goers happy. Many would be willing to pay more than the regular price to see a production in process.

Your Kitchen Away from Home

Sometimes you want to cook for yourself, the way only you can do it.

Imagine a chain of do-it-yourself restaurants. The restaurant provides the kitchen, the food, and the cooking utensils. You can cook for yourself or for you and a date or for a group, and eat in a pleasant environment.

Instead of *fast food,* this would be *slow food*, with lots of opportunity for conversation.

Ghost Restaurants in Paris and London

In Paris and London, about a third of the restaurants that looked interesting on Yelp and Google Maps turned out to be out of business. New businesses are quick to get listed, but failing businesses don't bother to unlist. Until these apps update their listings, they are inviting new competitors. In the meantime, it's far better to forget online recommendations and check the menus of nearby restaurants that you chance upon.

How to Facilitate Car Recalls

Car recalls are often ineffective. People with used cars never hear about them; and many who hear ignore them, even when safety is involved. The solution seems obvious. Most states require annual car inspections. Why not feed recall data into state inspection systems? Then when you go for an inspection, you could not only receive information about relevant recalls, but also the inspectors could ensure that the problems are fixed.

Print Obituaries in Large Type

Obituary notices are almost always printed in hard-to-read type, smaller than anything else in the newspaper. Considering that the elderly, many of whom have failing eyesight, frequently consult that part of the paper, that policy seems inconsiderate, even brutal. Newspapers could and should print death notices in larger type than the rest of the paper to make the information accessible by those who most need and want to read it.

Living Eulogies

I recently got a notice that a high school classmate had died. I'm 76. We're getting old. Such announcements are likely to become frequent, until there are few of us left.

I hadn't heard anything about him in 58 years. Now from eulogistic messages and obits, I get hints of a rich lifetime. I probably would have enjoyed knowing him.

I'm reminded of Antony's funeral speech in Julius Caesar. Antony ironically claims that the evil men do lives after them and the good oft lies buried with their bones. Eulogies are the reverse of that. They summarize a life, emphasizing the best. But they are only written and delivered when someone dies. That's a shame, and it's unnecessary.

I'd like to start a new tradition of *living eulogies.*

One possibility would be to hold a celebratory gathering (in person and remote) once every 20 years. Friends and family would gather and share their remembrances, saying what they would have said if this had been a funeral rather than a birthday.

If you were the centerpiece of such an event, you could pause and reflect. Is this the way you want to be remembered? What should you do to improve the narrative at the next such gathering? This would be an occasion where you could reconnect with people you had taken for granted or lost track of. It would also be a time to bond with the self you have been, considering how your acts, your words, your presence have affected others.

I'm reminded of a poem I wrote 52 years ago, on the occasion of the closing of a coffee house.

> Finnegan died,
> as people do every once in a while,
> so they held a funeral, an Irish funeral,
> and relatives and old friends who hadn't seen him for months or years all gathered,
> and it being winter, they held the picnic inside by candlelight;
> and everybody had such a good time
> that Granpa promised to die next year so they could have another good time just like it,
> and Granma volunteered for the next year,
> then all the aunts and uncles and cousins and third cousins and friends,
> till they had two centuries all booked up,
> and some pessimist in the crowd complained that he probably wouldn't live long enough for them to celebrate his funeral,
> and one of the aunts complained that hers was scheduled after one of the cousins, and she wasn't going to play second fiddle to any mere cousin;
> so Finnegan got up out of his coffin and told them to stop their squabbling —
> they'd just open a coffeehouse,
> and every week they'd close it again,
> and if people died, well, they could do it when they felt like it, in no particular order;
> but everybody could get together anyway, once or twice a week,
> and celebrate the funeral of the coffeehouse.

Advice for Rental Car Companies

In areas where snow emergencies are common, car rental companies are missing a major marketing opportunity. Customers should each receive an emergency package including:

- details on where to find the controls for windshield wipers, defrosters, etc.,
- a windshield scraper,
- instructions on how to return the car after hours, in case of a snow emergency, including what to do to get a ride home from the rental office during a snow emergency,
- a refund plan for occasions when a rental is cut short by bad weather or an option to extend the rental at no extra cost.

The Silence of the Cars

Have you ever started to cross the street and stopped short realizing that a soundless electric car was coming toward you? Have you ever started to change lanes and realized that an electric car was in your blind spot?

The silence of electric cars is dangerous — eliminating an important clue about what is near and how near and how fast approaching.

By national mandate, electric cars should make noises, as a safety feature to alert pedestrians and other drivers of imminent danger.

Better still, as such vehicles come to predominate, each should generate a unique sound. Then a vehicle could be identified not just by its license plate but also by its distinctive noise. A witness or a traffic video camera could record the sound, which could serve as an audio fingerprint, identifying the vehicle involved in a hit-and-run or other crime.

Outdated School Reunions

Bombarded with messages about my 50th college reunion, I wonder why reunions are tied to your year of graduation. When I think of my school experiences, I remember people in the class before and the class after, or who never graduated, those who were in the same major, or who took some of the same courses, or who were involved in some of the same activities.

High schools and colleges should organize reunions by schoolmates rather than classmates — people who attended school at the same time, rather than people in a single class.

Customer Service 101

Takis, a Greek immigrant, owned a Shell franchise in West Roxbury. Before I met him, I never paid attention to where I bought gas or who sold it to me. It was a commodity, and the people who did the pumping or handled the cash register were faceless. You might feel loyalty to a church or to an Elks Club, but one gas station was the same as another.

During the oil boycott of 1973, gas became scarce. When we needed gas, my wife and I took turns waiting in line for hours, and sometimes, despite the wait, got nothing. One day my wife got in line at Takis Shell, and the owner came out and told her, "You are a regular customer. You are special. We take care of you." We hadn't been *regular* customers before. But we were ever after that, benefiting from his special treatment during the boycott, and returning to his station and only his station for nearly 30 years.

When other gas stations automated so customers could pay at the pump and never see or talk to anyone, Takis taped over the credit card slots so you had to walk into his office to complete the transaction. He knew you by name, talked about himself and his family, and asked about you and your family; and he followed up on conversational details from one time to the next. Sometimes his wife or his son was behind the cash register. You weren't just buying gas; you were touching base with old friends. While in the office, you might buy snacks or ask for advice about a noise your car was making. And if your car needed fixing, that was where you took it for repair. Although you could have spread your buying power anonymously among a dozen local stations, you concentrated all your purchases there; and, without even considering price, you felt you benefited from this business relationship. Going to this gas station was like going to a general store a hundred years before — it was both social and business.

About 20 years ago, Shell put Takis out of business. Apparently, Shell decided to shut down franchises and instead to sell through company-owned stations. A few days before Takis turned over ownership, a reporter from the local newspaper was talking to him in his office when I happened to walk in. Takis pointed to me as an example of a loyal customer. And talking to the reporter, I realized how much I had learned from Takis over the years and that I had put those lessons into practice in building my Web-based business.

In the short run, automation may save you time and money; but in the long run, it could cost you dearly — making your operation the same as many others, putting you at the mercy of larger companies that sooner or later will price you out of business. It's well worth the time and trouble to talk to customers, to get to know them, to go out of your way for them. And if there's any kind of a glitch so you need to spend more time with a customer, that's an additional opportunity to build a relationship. The more what you have to sell is perceived as a commodity, the more important it is to build relationships with customers, to give them good reasons, both tangible and intangible, to return.

Today's company-owned station uses pay-at-the-pump automation. The air pump, which used to be free to all, is coin-operated. They do car washes but handle no repairs, not even flat tires. I don't know the people who work there. I never go there. Few people do. I miss Takis.

Internet Concierge

How much time do you waste trying to get help at Microsoft, Apple, Google, Paypal, Amazon, Dell, Hewlett-Packard, Verizon or other large Internet/computer/cellphone businesses? The instructions they provide use English words but inscrutably. They assume your screen matches theirs, when it doesn't. Hence you can't find the link you need. They insist you search for online solutions rather than letting you speak to live experts. They assume that you have a cellphone and a computer, both connected to the Internet.

Imagine an Internet concierge service. For a monthly fee, you have access to someone knowledgeable who can lead you through such nonsense or can go through the steps for you with remote access to your computer. That person could also give you reliable advice about upgrades that could simplify your online life.

Pain Simulator

Imagine a science fiction story that includes a *pain simulator*, a device developed for lesbian couples, allowing the non-pregnant partner to share the birthing experience. Such a product could be adapted for heterosexual couples, giving the husband the same opportunity to share. Using this invention, the non-birthing partner would feel the same pain, in the same regions of the body. The simulator could also serve as part of treatment for pain because the sympathy and togetherness of others helps ease the stress of someone in real pain.

Grandmother Bear

Imagine bear dolls dressed to look like grandparents so kids could play at hugging their grandparents. Grandparents would eagerly give such bears to their grandchildren. It would be natural to launch such a product around Grandparent Day — September 13.

Do-It-Yourself Alphabet Book

As a present for my first grandchild, Adela, when she was two years old, I assembled an alphabet book, with pictures of family members, friends, toys, etc. I included cousins, aunts, and uncles whom she saw rarely because they lived far away. The result was a time-capsule keepsake, a snapshot of the extended family at one point in time.

On the righthand page, a group photo appeared, with the question, "Where is ____?" Turn the page and you would see that person in closeup, with the words "Here is ____."

The solo shots were in alphabetical order. The number of pages per letter was unlimited, so no one was left out because a letter was *taken.*

The last group photo had the question, "Where is Adela?" When she turned the page, she saw a mirror with the caption "Here is Adela."

Fluff the Pillow

Suggestion for a simple videogame. No instructions. No words. Only touch for control.

The player sees an avatar and a pillow in an otherwise empty space. The avatar can touch, lift, toss, throw, pat, or fluff the pillow.

If and when the avatar lies on the pillow, if it is flat or dented, he looks sad. And if the pillow is fully fluffed when he lies on it, he looks happy. But when the avatar lies on the pillow for a while, the pillow once again becomes flat or dented. So the avatar has to fluff it again to become happy.

Independent Agencies for Bank Business?

Why does every bank have to have dozens, hundreds, even thousands of branch offices? From an ATM, you can do business with many different banks. Some of those ATMs are in those branch offices. Others sit in public places and are run by independent companies. Why not change banking regulations to allow independent agencies to handle all manner of consumer banking services for many different banks, obviating the need for the wasteful multiplication of single-bank branches.

Such agencies could also provide value-added services that no one provides today. Insurance agencies handle transactions for a variety of different insurance companies and can provide advice on which is the best one for you for each of your insurance needs. But you only deal with a single agent.

Similarly, banking agencies could make it easy to bank with multiple institutions, and yet have single-point access to all your accounts, face-to-face and also online.

Engineering Question — Windmills and Walls

In Boston, when the John Hancock Tower was erected near the Prudential, wind patterns changed: high winds blew out heavy windows from upper stories of the Hancock. With that in mind, I wonder if it is possible to deliberately change wind patterns and intensify wind flow by building walls.

For instance, when building an array of electricity-generating windmills in an open field, why not build walls perpendicular to the prevailing wind and angling inward, funnel-like? These walls could create wind-tunnel effects, intensifying the wind and focusing it on the blades of the windmills. Perhaps the walls could be movable, so they could be adjusted to maximize the effect at different times of the day and of the year.

How to Cope with Reduced Weight on the Moon

China, Japan, India, and the US are all now racing to put men on the moon. Unlike the first time around, the objective is to establish permanent bases. That raises the question of how well humans can cope long-term in the moon's reduced gravity. We evolved on Earth to live in Earth gravity. A long-term change in gravity would affect bone mass, muscle tone, and the function of internal organs.

The moon's gravity is about 1/6 that of Earth. So someone who weighs 180 lbs. on Earth would weigh just 30 lbs. on the Moon. In science fiction, the usual solution for weightlessness is magnetic boots or space ships and space stations that spin, so centripetal force substitutes for gravity. But those solutions would not make sense on the surface of the Moon.

Why not just add weight? For instance, a 180 lb. astronaut could wear a suit weighing 900 lbs., bringing the total weight to 1080 lbs. on Earth and 180 on the moon.

The extra weight need not be dead weight. It could include oxygen, water, and food; armor to protect against hazards such as meteorites, high-speed space debris, and cosmic rays; and high-tech gear that extends a person's capabilities.

Cigarettes — why not put a filter on the other end?

Today's cigarette filters are on the receiving end — between the cigarette and the mouth of the smoker. But much of the bad effects of smoking — to the smoker and to those around — are from the smoke that comes out the other end.

So why not design a filter for the burning end, a reusable plastic or metal gadget to put over the cigarette? It would allow air to enter so the tobacco could burn. And it would filter out harmful substances before releasing smoke into the air. Such a device could eliminate the odor of smoke as well. And it could also shut down after a period of inactivity, extinguishing the burning, as a way to prevent fires.

If such gadgets existed, it would make sense to allow people who use them to smoke in public places, since their use would eliminate the danger and even the odor of second-hand smoke.

Why not save a hair?

In TV mystery shows that focus on crime and forensic evidence, medical examiners sometimes exhume bodies to verify the identity of the deceased. Sometimes their efforts at identification are stymied by the fact that the body was cremated.

Why not take a hair from everyone who dies, and securely store that hair in case the need ever arises to test the DNA? That's a simple, non-invasive procedure that could obviate the need for costly and emotion-wrenching exhumation. As for privacy concerns, the person is dead. And the DNA test is only performed if ordered by a judge when there's reasonable cause.

Objections? Perhaps some would see this as a first step toward a national identity system based on DNA tests performed on newborns. But would that be bad?

Odor and Identity: Perfume as Infidelity Training. I Stink Therefore I Am

At a drug store this morning, the clerk at the counter accidentally spilled perfume on her hands. She could do nothing to get rid of the smell. She'd have to wait for it to wear off.

That's when it occurred to me that there should be perfume-removal products. We have removal products for nail-polish and makeup. Why not for perfume?

Such a product could be a winner in and of itself, but it also could be a boon to the perfume industry. If you could quickly, easily, and effectively remove a fragrance, you could get into the habit of changing fragrances, as often as you change your clothes or your makeup. There could be morning, afternoon, evening, and bedtime fragrances. There could be fragrances for business, for home, for shopping, for social occasions. Advertising could convince consumers that such changes are a social necessity, that you'd feel naked without a fragrance, and gauche using the wrong perfume, just as you would wearing the wrong kind of clothes or the wrong kind of makeup.

Then it occurred to me that perfume — not products that clean the body and remove distasteful odors, but rather products meant to generate an artificial odor — have a negative side-effect. Normally, the touch, the sight, the taste, and the smell of your partner are all connected with your partner's identity. In the process of falling in love, like Pavlov's dog, you learn to associate all those sensory cues with your beloved.

By nature, a person's natural scent is unique. But commercial perfumes are mass produced. If a woman wears a perfume when she's with her partner, she unwittingly trains her partner to associate that scent, rather than her own, with romantic feelings. She exchanges a characteristic that can uniquely identify her for something any other woman could readily imitate. You might say she trains her partner for infidelity.

Intelligence Test Wish List

We need to identify individuals with talent for finding unusual, yet workable, solutions to practical problems.

I believe that psychologists should use videogames to test intelligence. And rather than write new ones, they should analyze current, popular games to correlate performance with intelligence and out-of-the-box thinking. Today's kids are far more comfortable playing videogames than performing the pencil-and-paper or face-to-face tests designed generations ago to measure IQ. Without abandoning the old tests, we should add these new ways of measuring intelligence, to identify talent that might otherwise be overlooked.

Pee Pill

There are medications to help people with conditions that lead to sudden, intense urinary urges and leakage. But there does not seem to be anything available for ordinary people who find themselves in unusual circumstances that make it awkward, difficult, or impossible to urinate.

We need a pill that temporarily reduces the urge to urinate. Take it before going to a performance or a movie theater. Take it before you participate in a sporting event. Take it when you might be caught for hours in a traffic jam.

Programmable Mattresses

Some people feel inclined to shift from one bed or sofa to another in the middle of the night. I suspect that different mattresses are more compatible with different stages of sleep — e.g., the difference between deep sleep and dreaming or REM sleep.

Mattress companies now sell adjustable mattresses such that two people can set their separate sides to their preferred firmness. It would make sense for them to make such mattresses programmable, so users could choose to change the setting at pre-set times or to have the bed sense changes in sleep mode and adjust accordingly.

Infocard

Imagine using a cellphone app for the exchange of personal information. You select the level of information you want to share. For instance,

- business
- friend
- acquaintance

For each level, you predefine the kinds of information you want to provide, and you pre-enter that information. The exchange can happen in person or at a distance, perhaps even during a remote meeting like Zoom.

You turn on *handshake mode* and invite the other party. If the two of you agree with the level of contact, the info is exchanged.

In person, you could point the camera of your phone at the target person to identify *infocard* members by facial recognition and to initiate a handshake. And you could choose if and when to be open for recognition and for handshake requests.

This app could also have levels for flirting, dating, etc. And the contact/exchange could take place discretely, with no sound to indicate that it was happening. Info regarding who you have contacted and at what level and what information was exchanged could be kept private.

Dog-Talk App

Imagine an app intended to help you meet interesting people walking their dog(s). With a camera shot, the app recognizes the breed, estimates the dog's age, and provides anecdotal information about that breed. The app itself could be a conversation starter.

This same app could be useful for identifying lost and stolen dogs, including ones for which a reward is being offered.

Would You Like a Huggle?

How many times have you needed a hug from someone who is far away? How many times has someone sent you a message that ends with *hugs*?

I imagine a cuddly stuffed toy, a *Huggle,* available in a variety of shapes and sizes, with an Internet connection and the ability to move on command. You give it to someone you love to keep it like a pillow on a bed or sofa. Then when the mood strikes, you can trigger a remote hug.

Alerted through a cellphone app that a hug is being sent, your lover clicks to accept it or postpones delivery to a more appropriate time and place. Only you (the gift giver) can activate Huggle with the app. And you can customize Huggle's motions to match the mood of sender and recipient.

Party Line

When I was a kid, I lived in Plymouth, NH, population less than 3000. The town printed a calendar, and you could pay 25 cents to have your birthday or anniversary printed on it so everybody would know.

In those days, your telephone line was a party line, shared with about a dozen other families. If you had an emergency, the others would get off so you could take care of it. But normally, you had to wait until others had finished talking before you could place your call. And you might listen in to know when they were done. And you might join in the conversation, the other parties all being friends and neighbors.

Yes, there was a lack of privacy with those party lines, but there was plenty of community. I'm surprised that no one now offers such a service over cellphones.

Imagine — you and your friends subscribe as a party, maybe a dozen people. When one of you calls another, the rest are alerted that the call is going on so they can connect if they wish and listen in and participate. Everyone connected sees who started the call and with whom and who else joined in. The call ends when the last person hangs up. (It could also be done with video as an option).

Rear-View Mirror

How can a woman know what she looks like when seen from behind? She doesn't see her back in the mirror, not even in two mirrors, one behind the other, because she herself is in the way. But if she cares how she looks and what others think of how she looks, she would want to check how she looks from behind.

Why not an extendable selfie-stick with a swivel, so you can take photos or videos of your back from 1–5 feet away? And why not connect that with your laptop or tablet, so you can see yourself live and in motion — to know what you look like walking or jogging or exercising?

Clothing stores could offer that gadget to help you choose what to wear and how to wear it. At home, it would help you choose outfits for particular occasions.

And if your significant other isn't nearby, you could share such images of yourself.

Invisible Leash for Toddlers and Dogs

(This is wishful thinking. I don't know how it could be accomplished. It feels like a step beyond invisible fences and fancy leashes.)

I can imagine a harness or coat electronically connected to your cellphone. If a pet or child strays near the limits you have set, you can pull them back, remotely. This device should not inflict pain (like invisible fences). Rather, obedience should trigger pleasurable feelings. And it's not just for defining your yard, but also can be used when you go on walks.

Sometimes you have too much to do. Your dog fetched the leash and dropped it at your feet, with a hopeful look. Or your toddler is whining to go outside. But right now is not a good time for you. In that case, you should be able to set GPS-based parameters, defining an area that you consider safe, such as your back yard or block. And part of the reward/punishment for not crossing the limits could include the sounding of your recorded voice — approving or warning (in crescendo, if necessary).

Rape Recorder

The chastity belt of the Middle Ages was not intended to control a wife, to prevent her from having sex with other men during the husband's absence. Rather, it was meant to protect her from rape when he wasn't around.

I'm surprised that we don't have a modern equivalent. Multiple devices and services are available to protect homes. Why not personal protection using similar technology?

I imagine an electronic device triggered by a violent assault.
— an alarm to scare the assailant away,
— automatic notification of police, with GPS location, and
— automatic recording of voice and video, and transmission of same to police.

In a just world, such devices, with related services, would be free for residents of high crime areas.

Testicle Warmer as Prophylactic?

Sperm are sensitive to high temperatures. Many web sites provide advice on how to avoid warming the testicles, in an effort to enhance male fertility, to boost the likelihood of pregnancy.

As noted at the web site of Fertility Associates of Memphis, "Testicles are located outside of the human body because the ideal temperature for sperm production, and thus for male fertility, is about three to four degrees below the ideal human body temperature of 98.6°F."

To what degree is raising the temperture of testicles effective in preventing pregnancy? And is infertility due to testicle temperature temporary and reversible?

There are disposable hand warmers and foot warmers that provide 9–18 hours of warmth. You slip them into gloves, boots, pockets, hard hats, or between layers of clothing. Why not testicle warmers, using similar chemicals or a different technique? There might be a way to reliably generate enough heat to render sperm incapable of impregnating for a predictable period of time, but without permanent damage. Depending on the level of reliability, this could be used instead of or in addition to female contraception.

I am reminded of the codpieces worn in Shakespeare's day. Could they have had a testicle warming effect? Perhaps the kind of cloth and the degree of openness could have helped control fertility, whether intentionally or not. I can imagine Shakespeare using wool to wrap himself tightly for hours before visiting his dark mistress.

Rainbow Generator

Rainbows can appear in the sky when there are water droplets in the air and sunlight shines from behind them at a low angle. This happens at a rare moment when the sun is out but it's also raining, and you happen to be standing in the right place to see it.

Rainbows also can be generated artificially, when you hold a prism at the right angle on a sunny day or when sunlight strikes the spray from a hose or a fountain in just the right way. So it should be possible to generate rainbows predictably and automatically.

I can imagine a cellphone app that communicates with an intelligent fountain or hose. The app determines if and when the position and brightness of the sun makes rainbow creation feasible. Then you turn on the water spray, and the app prompts you to move closer or farther, left or right, to the best position for viewing the rainbow.

Play Chess with the Dead

(What I am suggesting here would be very easy to do. If some company already offers such a service, please point me to it. If not, I'm hoping this will inspire someone to do it.)

Powerful chess-playing programs are inexpensive or free. My favorite is Shredder, which is available as an app on the iPhone and probably in other forms. These programs adapt to your strength, so you have a reasonable chance of winning, and automatically change as your strength improves or declines. There are also programs that include games played by champions and grandmasters, so you can watch those games played out on the screen, sometimes with commentary.

I would like a program that can analyze a player's style and build a chess personality, such that the computer can play like a certain individual. In other words, the computer's decision-making algorithm would be weighted with the proclivities of a particular person. You could choose to play a ghost match against Lasker or Capablanca, Alekhine or Fischer. Or you could train for a tournament game against a particular rival.

To enable the system to build a chess personality, you would enter the moves that person played in previous games. For historic games, those are probably already stored as part of the program. This would be a different way of using that data. For contemporary players — frequent tournament competitors or friends you often play for the fun of it — you would need to enter the moves of a significant number of their games (maybe 100 games as black and 100 as white). The company providing this service should test to determine the optimum number for establishing a realistic profile.

With practice, you could improve your chances of winning against such individuals. You also could enter your own games so the computer could play as you against you, allowing you to explore your vulnerabilities and find better follow-on moves for your favorite lines of play.

And if and when a friend isn't available, you could continue to play against the simulated version of him or her.

And you could program such a system to continue to match you and your friend after your friend can no longer play— a peculiar form of immortality, a stopgap, until it's possible to play in the afterlife. (What is Heaven for except the fulfillment of impossible wishes?)

Part 9:
Everyday Life — How To Live, How To Cope

Exemplary People: The Machinist and the Postal Clerk

(I only knew these people briefly and in a single aspect of their lives, but their every action seemed to indicate focus, intensity, consistency, and dedication. They were proud of what they could do, and how they did it showed who they were.)

The Machinist

Bob was foreman at Econotool, a machine shop in the suburbs of Philadelphia. I worked there for a couple of summers when I was in college. They made cutting tools which were resold by Black and Decker. They taught me a few simple repetitive tasks. I silver-soldered carbide blades to steel shanks, keeping a close watch on temperature and time. I also ground those carbide blades on a diamond wheel, using a rig designed to sharpen them at the specified cutting angle. It was easy to clamp the raw piece into position and grind away. But it was difficult to do this hundreds of times a day and give it your full attention — to not fall half-asleep with boredom, and blunder disastrously.

Bob knew everything there was to know about the machines I was working on and all the other machines in the plant. He could clean them and fix them. Given a blueprint, he could make rigs, set them up, and adjust them, so they would turn out quality product repeatedly. If he didn't have a replacement part, he could machine that part from scrap metal; and he enjoyed such unexpected challenges.

This was blue-collar work, but Bob wore a brightly clean white shirt, proudly. He worked hard, but at the end of the day that shirt was always as clean as at the beginning. He worked with no waste movements, no spills, no accidents — all from unremitting dedication to what he did.

He reminded me of a craftsman in the Middle Ages whose craft was his identity and became his name, like Cooper or Smith, who conscientiously devoted his full attention to his work, day after day. From the way Bob worked, you knew he believed that what he did mattered, and that doing it well gave his life meaning. Money was secondary. His work was his religion. He handled his tools and his machinery with the respect and pride of a priest serving mass.

The Postal Clerk

For 30 years, Jack, now retired, was a clerk at the Post Office in West Roxbury, MA. He approached his job with a level of seriousness and respect similar to Bob at Econotool. He knew every postal regulation by word and in spirit. When a question arose, his fellow clerks turned first to him, not the book. When a customer had a question or misunderstood the options and was about to make a costly decision, Jack explained the rules and also the practical aspects of how the mail is handled, without talking down. You got a lesson on the Postal Service and also on life, from someone who could have been a great teacher but who took pride in being a great postal clerk.

Jack didn't just sell you stamps and make sure you had filled out the right paperwork for shipping packages overseas or applying for passports. While doing everything that needed to be done as efficiently as it could be done, he would smile and give you a tidbit of information or advice for the future; and he'd often set in motion some of the many elaborate gadgets he had on display in his work area. These were liquid mind-mystifiers — turn them upside down and bubbles moved in random but beautiful patterns, confined yet free, and demanding your attention.

I hope that in retirement Jack still finds ways to take pleasure from and pride in the necessary details of life. I miss him.

Procrastination Index

At the start of the day, identify the one thing likely to be hardest for you to get around to. That's your procrastination index. For me that's usually calisthenics. If you do that one thing early, you're going to have a high-energy, creative, productive day. If not, you might as well roll over and get more sleep.

In any case, believing that principle will give you incentive to do that one thing.

Creative Procrastination

Working in an office, we have to deal with the expectations of bosses and co-workers. Our social and physical environments provide reminders of what needs to be done and when to do it. Working at home, alone, it's easy to procrastinate or to shuffle from one project to another, not making substantial progress on any of them. It sometimes takes an effort to motivate ourselves to get things done on time.

I use *creative procrastination* to overcome that difficulty.

At any given moment there are dozens of things I could and should be doing. I list and prioritize them. Many times, the most pressing task is something I don't want to do right now. I'd rather do something else. So, I work on that something else. It's also on the list, and I'd have to do it sooner or later.

While doing that *something else*, I keep remembering the task I put off that I really should be doing. That feeling of guilt generates energy, motivating me to do the alternate task faster and better.

Eventually, I can't put off the high priority task I'm avoiding. I need to start it pronto or I won't be able to get it done on time and that failure will have consequences. In that case, I need to think of another task that I will eventually need to do and that is even more undesirable than the high priority one. I try to convince myself that that task is important and urgent. The more I think about the substitute task, the more attractive the priority one appears. So I get it done.

Over time, we need to discover our natural rhythms, categorizing the things we need to do on a regular basis, and getting a feel for how often we get the urge to do such things. For instance, Web site updates, paying bills, cleaning the house or yard, and doing creative work. For me, the cycle is about a week. If I try to pay bills and balance my check book on a day when I feel like working on a creative project, that's like pushing rocks uphill. Likewise, working on a creative project when my mind would prefer the relaxing tedium of a repetitive task, is laborious and unproductive. So, I try to do the things I need to do when it feels natural to do them, modulating that with creative procrastination, as described above.

Once we find our natural rhythm, we need to schedule, loosely. For instance, say my rhythm is a week or two and one of the chores is finances. In that case, I would aim to deal with finances on the same day every week and try to make that a habit. But if the right mood isn't there that day or something else comes up, I don't worry about it. Over the course of a week or two, I should be able to cover what needs to be covered, by just following my natural inclinations. And since I'm doing this work when I'm in the right frame of mind, I'm able to work faster and better.

If there's some regularly recurring task that needs to get done but that I never have the urge to do, even using creative procrastination, then I try to make adjustments so it's no longer necessary for me to do it, or I hire someone to do it for me. I am who I am. Why fight it?

Done List Instead of To-Do List

When I worked alone at home for my ebook publishing company, I had no boss, no colleagues. My *to-do list* was growing out of control. Every day it got longer and more daunting. There was no way I could keep pace, much less catch up. The list was intended as a way for me to prioritize and get organized, but it became a chore in and of itself.

Then I shifted from *to-do list* to *done list*. That change improved my state of mind and productivity.

My *done list* recorded what I actually did each day. I categorized what I did into four main areas: family, maintenance, business, and personal. And each of those categories had subcategories. For instance, maintenance included health, food, and house. Personal included writing, reading, languages, TV, and other entertainment.

Before, time seemed to disappear. The day would end, and I would wonder how I managed to waste so much time. My focus would be on all the *to-do* list tasks that I hadn't gotten to. With my *done list*, I saw reminders of what I had accomplished and enjoyed. And the categories gave me a feel for what I actually did with my time, showing my patterns, making sense out of what seemed to be random.

Of course, I still kept a short, manageable list of things I planned and expected to do that day. But the focus was on the *done list*, and adding a new item felt good.

While my circumstances are different now, I still keep a *done list*, to which I can now add writing this essay.

How to Expand Time

Step One - Get Moving

Do one small thing a day that could eventually make your life a little better. Don't try to come up with a master plan. Don't try to make sense of everything. Just focus on small activities that will move you forward. At some point, it might dawn on you that these actions are taking you in a particular direction. Then you may make course corrections to advance more quickly. But first, get moving.

Step Two - Exer-Work

As you get older, everything becomes harder and takes longer. Increasingly you need physical exercise for health, which leads to a time crunch — how can you spend an hour a day walking or jogging when there isn't enough time in the day to do what you want and need to do?

To deal with this problem, I categorized my typical activities as maintenance, relaxation, exercise, or progress, and I kept track of when I did what. Then I modified how I characterized activities so they could do double duty, getting done what I needed to do and benefitting from the exercise of doing it.

Remember, God invented snow to shovel, grass to cut, and leaves to rake so you would be motivated to exercise. He also created dogs so you'd have good reason to walk.

Writing is part of who I am and what I do. So, I begin working on a project before going out to exercise. Then ideas occur to me naturally as I walk or jog. I bring along a notebook. And instead of the exercise being a boring chore, I enjoy it; and by the time I get home, I'm anxious to start writing again.

Step Three - You Don't Need to Finish Everything at Once

Today's computers are *digital*. They are based on only two possibilities — yes or no, on or off. But people are by nature *analog*, with a continuous range of possibilities. For us, little changes matter because they accumulate and take us where we want to go a little bit at a time.

Tangible Goods vs. Experiences

When I was four years old, my parents gave me ten cents a week as an allowance, and I could walk to the corner drug store and buy what I wanted.

The big choice was between tangible objects and ephemeral experiences. If I bought plastic cowboys and Indians, I could play with them over and over. And if I bought comic books, I could read them over and over. And in either case I could share or trade them with friends.

If instead, I bought penny candy or ice cream, while the experience would be delightful, it wouldn't last. The act of enjoying it would destroy it. I would end up with nothing. Despite temptation, I almost never opted for candy or ice cream.

When my kids were young, I hoped that they would enjoy the toys and comic books that I had enjoyed at their age. But they showed little interest in the old stuff, caught up as they were in the new objects of popular culture that they and their friends saw advertised.

Now I value experiences. The boxes of plastic cowboys and Indians and comic books just take up space. They're a burden to me, and, after I go, they'll be considered trash. At this stage of my life, I want to enjoy the moment rather than accumulate tangible goods. I value ephemeral beauty and ephemeral pleasure, and memories of such moments.

How to Fall Asleep Naturally

(This works for me. Perhaps it will for you.)

Whatever is on your mind — things to do, things to remember — write them down and put the list aside.

Lie on your back. Stretch out. Separate your legs so they aren't touching one another. Put your arms under the covers, not touching your body.

From toes on up, make whatever adjustments are necessary to make yourself relaxed and comfortable.

Empty your mind. Think of nothing. Absolutely nothing.

If, after a few minutes, you are still awake, count slowly, thinking of nothing but those numbers.

If you become aware of any discomfort in any part of your body, relax that part and start counting again.

If you believe this will work, it will probably work. Anxiety that you won't be able to fall asleep is a barrier to falling asleep.

If you are still awake when you reach a hundred, this method probably won't work for you.

Stopping at a Stoplight in the Middle of Nowhere

When the light turns red, I stop. It doesn’t matter whether there’s anyone around to see. It doesn’t matter whether there's a videocam to record the moment. I stop as an affirmation of a social contract of mutual concern and responsibility. That's what we all should do, regardless of whether we might get caught. I want to have faith that others won’t cut corners, take unnecessary risks, and cause avoidable accidents. Waiting for a couple minutes at a stop light in an empty intersection in the middle of the night, I feel a connection with other drivers I've never met who would do the same for the same reason.

I’m reminded of the old question, "If a tree falls in the forest with no one to hear it, does it make a sound?" Does there have to be a witness for an act to matter? I believe we are all connected to one another by what we do and what we don’t do, whether alone or in the presence of others — connected every moment of our lives.

---

written from 1995 in West Roxbury, MA to November 2022 in Milford, CT

CPSIA information can be obtained
at www.ICGtesting.com
Printed in the USA
LVHW040105060123
736479LV00011B/877

9 781958 878477